SELECTING THERMOPLASTICS FOR ENGINEERING APPLICATIONS

PLASTICS ENGINEERING

1. Plastics Waste: Recovery of Economic Value, *Jacob Leidner*
2. Polyester Molding Compounds, *Robert Burns*
3. Carbon Black-Polymer Composites: The Physics of Electrically Conducting Composites, *edited by Enid Keil Sichel*
4. The Strength and Stiffness of Polymers, *edited by Anagnostis E. Zachariades and Roger S. Porter*
5. Selecting Thermoplastics for Engineering Applications, *Charles P. MacDermott*
6. Engineering with Rigid PVC: Processability and Applications, *edited by I. Luis Gomez*
7. Computer-Aided Design of Polymers and Composites, *D. H. Kaelble*
8. Engineering Thermoplastics: Properties and Applications, *edited by James M. Margolis*
9. Structural Foam: A Purchasing and Design Guide, *Bruce C. Wendle*
10. Plastics in Architecture: A Guide to Acrylic and Polycarbonate, *Ralph Montella*
11. Metal-Filled Polymers: Properties and Applications, *edited by Swapan K. Bhattacharya*
12. Plastics Technology Handbook, *Manas Chanda and Salil K. Roy*
13. Reaction Injection Molding Machinery and Processes, *F. Melvin Sweeney*
14. Practical Thermoforming: Principles and Applications, *John Florian*
15. Injection and Compression Molding Fundamentals, *edited by Avraam I. Isayev*
16. Polymer Mixing and Extrusion Technology, *Nicholas P. Cheremisinoff*
17. High Modulus Polymers: Approaches to Design and Development, *edited by Anagnostis E. Zachariades and Roger S. Porter*
18. Corrosion-Resistant Plastic Composites in Chemical Plant Design, *John H. Mallinson*
19. Handbook of Elastomers: New Developments and Technology, *edited by Anil K. Bhowmick and Howard L. Stephens*

20. Rubber Compounding: Principles, Materials, and Techniques, *Fred W. Barlow*
21. Thermoplastic Polymer Additives: Theory and Practice, *edited by John T. Lutz, Jr.*
22. Emulsion Polymer Technology, *Robert D. Athey, Jr.*
23. Mixing in Polymer Processing, *edited by Chris Rauwendaal*
24. Handbook of Polymer Synthesis, Parts A and B, *edited by Hans R. Kricheldorf*
25. Computational Modeling of Polymers, *edited by Jozef Bicerano*
26. Plastics Technology Handbook: Second Edition, Revised and Expanded, *Manas Chanda and Salil K. Roy*
27. Prediction of Polymer Properties, *Jozef Bicerano*
28. Ferroelectric Polymers: Chemistry, Physics, and Applications, *edited by Hari Singh Nalwa*
29. Degradable Polymers, Recycling, and Plastics Waste Management, *edited by Ann-Christine Albertsson and Samuel J. Huang*
30. Polymer Toughening, *edited by Charles B. Arends*
31. Handbook of Applied Polymer Processing Technology, *edited by Nicholas P. Cheremisinoff and Paul N. Cheremisinoff*
32. Diffusion in Polymers, *edited by P. Neogi*
33. Polymer Devolatilization, *edited by Ramon J. Albalak*
34. Anionic Polymerization: Principles and Practical Applications, *Henry L. Hsieh and Roderic P. Quirk*
35. Cationic Polymerizations: Mechanisms, Synthesis, and Applications, *edited by Krzysztof Matyjaszewski*
36. Polyimides: Fundamentals and Applications, *edited by Malay K. Ghosh and K. L. Mittal*
37. Thermoplastic Melt Rheology and Processing, *A. V. Shenoy and D. R. Saini*
38. Prediction of Polymer Properties: Second Edition, Revised and Expanded, *Jozef Bicerano*
39. Practical Thermoforming: Principles and Applications, Second Edition, Revised and Expanded, *John Florian*
40. Macromolecular Design of Polymeric Materials, *edited by Koichi Hatada, Tatsuki Kitayama, and Otto Vogl*
41. Handbook of Thermoplastics, *edited by Olagoke Olabisi*
42. Selecting Thermoplastics for Engineering Applications: Second Edition, Revised and Expanded, *Charles P. MacDermott and Aroon V. Shenoy*

Additional Volumes in Preparation

Preface

Most of us will agree that the task of selection in any endeavor is easiest when there is a limited choice—or better still no choice at all. The larger the number of options, the more variables that must be considered in the process of narrowing the choices to make a final selection. When it comes to selecting a proper thermoplastic for a particular engineering application, the task is extremely difficult because the choices are numerous; furthermore, the number of choices is continuously growing with the introduction of newer and modified resins.

One often wonders whether selecting thermoplastics for engineering applications is an art or a science. If it is an art, then it has to be based on individual skill, imagination, and practical experience. On the other hand, if it is a science, then the selection process must involve fundamental theory, systematic experimentation, disciplined research, and logical procedures. We believe that the resin selection process must have its foundations in a combination of both art and science. A good balance between the two will increase the chances of success and alleviate the hazards of guesswork.

Since new applications are envisaged every day, decisions for selecting the proper thermoplastics might be assumed to be a fairly routine and repetitive exercise. In reality, this is not the case. When nearly 50 people who are experienced in all aspects of resin selection were interviewed, it was found that no two of them could agree on a logical process for selection. This underscored the need for the present book.

Ideally, one would develop a procedure for resin selection by assembling a number of actual developments in engineering plastics from material suppliers, molders, and end users, and arranging them logically based on practical case histories. It was found, however, that case histories are not readily available, records are not always kept meticulously, and most often selection decisions are made based on pragmatic rather than logical considerations.

For these reasons, it proved very difficult to outline a single thermoplastic selection procedure for all engineering applications. Nevertheless, we found that we could provide guidelines to channel the thinking process in order to make the task easier. While we do not present an exhaustive list of all available thermoplastics and their properties, we consider some of the important representative thermoplastics in order to demonstrate the logical sequence of thought needed to make an informed selection of the right thermoplastic for a given engineering application. We also demonstrate how thermoplastics can be ranked based on their usefulness criteria and show how a computerized approach may be tried for resin selection using these rankings.

A step-by-step procedure is given for using an expert system demonstration program named SELECTHER, which guides the end user in *selec*ting *ther*moplastics in prioritized order of usefulness for a particular envisaged engineering application. Interested readers may revamp the program to include a different list of candidate resins or expand it into a full-fledged version that can help select from among a larger number of thermoplastics.

We firmly believe that by applying the principles discussed in this book with a logical scientific approach, the task of selecting the right thermoplastic for a particular engineering application can be made much easier and the selection more appropriate.

Charles P. MacDermott
Aroon V. Shenoy

Acknowledgments

Physical property data used in this book were taken from a number of sources, with a large share of them from promotional literature available from the major material suppliers. We are particularly indebted to the following companies for the information provided by the bulletins listed below:

Borg–Warner Chemicals
Injection Molding Polymers—Grade Selection Guide, Cycolac ABS
E. I. du Pont de Nemours & Company
Zytel® Nylon Resin—Design Handbook
Delrin® Acetal Resin—Design Handbook
Zytel ST Super Tough Nylon Resin
Designing the Car of the Eighties
A Guide to Standard Physical Tests for Plastics, P. N. Richardson
Designing with du Pont Engineering Plastics
Delrin Acetal Resins—Molding Manual
Rynite® Thermoplastic Polyester Resin—General Guide to Products and Properties

Minlon®—Design Handbook
Zytel Glass Reinforced Nylon Resins—Molding Manual
Plastics: An Overview, by David Stotz
Celanese Plastics and Specialties Company
Celanex® Thermoplastic Polyester—Properties and Processing
Celanese Engineering Resins
Celcon® Acetal Polymer
Celanese Nylon Selector
Celanese Nylon—Glass Reinforced Nylon, Bulletin N 1B
Dynamit Nobel
Dyflor® 2000—Polyvinylidene fluoride
General Electric Company
Lexan®, A Good Name to Stand On
Lexan Polycarbonate Resin
Designing with Lexan
Valor® Resin—Injection Molding
Noryl® Engineering Plastics
Hercules Company
Selection Guide to Mineral Filled Profax® Polypropylene
Union Carbide
Udel® Polysulfone Product Data
Designing for Polysulfone
Ardel® Polyarylate

Contents

SELECTING THERMOPLASTICS FOR ENGINEERING APPLICATIONS

1

Definition of Engineering Thermoplastic Resins

INTRODUCTION

Plastics have touched every important aspect of our life in terms of various human needs: food (packaging), water (ion-exchange membranes for desalination), clothing (synthetic fibers), shelter (laminates and paints), transport (automobile and airplane body components), health (artificial organs and surgical instruments), entertainment (TV cabinets and cassette casings), office equipments (computer casings and furniture), travel (luggage), and so on. Plastics are increasingly replacing conventional materials such as glass, wood, metal, and ceramic in engineering applications. For example, corrosion-resistant coatings of Teflon® and polyphenylene sulfide are used in chemical industries in place of glass lining; photosensitive plastics are used for making microelectronic circuits; metal and ceramic pipes have been replaced by poly(vinyl chloride) pipes in agricultural applications, and so on.

The use of plastics has expanded explosively over the past 50 years, to the point that today in the entire world, millions of metric tons of plastics are produced annually. A large part of this huge volume is extruded into film, sheet, pipe, wire and cable coatings, and tubing. Another significant share is injection or blow molded into toys or into throwaway articles such as bottles or food packages. Most of the plastics used for these purposes are considered general-purpose materials or commodity plastics and are not covered as part of the discussion in this book. In terms of cost, commodity plastics are the cheapest of the thermoplastics. They are produced in the largest quantities and are used in a maximum number of applications.

A smaller share of the total plastics production, although still measured in the billions of pounds annually, is reserved for "engineering thermoplastics." These are high-performance materials[1] that provide a combination of high ratings for mechanical, thermal, electrical, and chemical properties. They are capable of withstanding high loadings for long periods of time at elevated temperatures and in adverse environments, and behave in a predictable manner when subjected to design techniques and formulas. They find application in various automotive, structural, and industrial markets. Typical applications include electrical switches, gears, bearings, cams, auto ignitions, etc. The following plastics are generally categorized as engineering thermoplastics, as shown in Table 1.1: acrylics, polyacetals, nylons, thermoplastic polyesters, polycarbonate, polysulfone, polyphenylene sulfide, polyphenylene oxide/polystyrene blends, polyimides, polyamide-imides, polyethers, fluoropolymers, and so on.

At times, commodity plastics such as high-density polyethylene (HDPE) and polypropylene (PP) can be utilized in engineering applications when they are filled with the right type of reinforcing fillers.[2–5] Fillers are also used in conventional engineering thermoplastics for cost reduction or to impart specific functional properties.[4–9] A variety of functional fillers such as aramid/carbon fibers for reinforcement,[10,11] metallic coated glass fibers for electrical conduction,[12,13] barium or strontium ferrites for magnetic properties[14,15] are increasingly being used in plastics. The developments in filled plastics have been discussed in a number of books,[16–21] and a list of the variety of fillers that have been tried is also available.[22] In general, particulate-type fillers, such as wood flour, calcium carbonate, clay, and sand, are used as cost reducers or extenders; whereas fibrous fillers (such as wollastonite, glass fibers, and Franklin fiber) and platelike fillers (such as mica) represent reinforcing additions that improve the mechanical properties of the plastic.

Commodity plastics such as polystyrene (PS) and poly(vinyl chloride)

TABLE 1.1 Selected Thermoplastics Used in Engineering Applications

Engineering thermoplastics					
		Modified commodity plastics			
Conventional	Specialty	Blends	Composites	Blends	Composites
Acrylonitrile butadiene styrene (ABS)	Fluorocarbon copolymer (Teflon® PFA)	(e.g., ABS/PVC, PVC/acrylic, CPE/PVC)	(e.g., talc-filled PP, mica-filled PP)	ABS/PC	Mineral-reinforced 6/6 nylon
Acetal	Polyether sulfone (PES)			ABS/polysulfone	Glass-reinforced 6/6 nylon
Polyamide (e.g., 6 nylon, 6/6 nylon)	Polyarylate (PAr)			SAN/polysulfone	Glass-reinforced PC
Polycarbonate (PC)	Polyphenylene sulfide (PPS)				Glass-reinforced PPE-based resin
Polyphenylene oxide (PPE-based resin)	Polyamide-imide (PAI)				Glass-reinforced PET
Polysulfone	Polyether-imide (PEI)				Glass-reinforced PBT
Thermoplastic polyester (e.g., PBT, PET)	Polyether ether ketone (PEEK)				

(PVC) can enter into engineering applications via blending, wherein they are physically mixed with certain engineering thermoplastics for improved performance. There are a number of commercial thermoplastic blends[23] in use, as summarized in Table 1.2. It should be noted that two engineering thermoplastics can be combined to form a useful polymer blend, as in the case of ABS/PC, ABS/polysulfone, or SAN/polysulfone. In polymer blends, the individual polymers are chemically different and do not form covalent bonds as in copolymers such as SAN or terpolymers such as ABS.

In the present context, the words "engineering thermoplastics" encompass all thermoplastics which can be utilized in engineering applications. By this connotation, all homopolymers and copolymers listed under the engineering thermoplastic category in Table 1.1, along with all commodity plastics that are appropriately filled with reinforcing or functional fillers, or blended appropriately for improved performance, would all be termed as engineering thermoplastics. Since the specialty plastics shown in Table 1.1 also go into engineering applications,[24,25] they too are included in this broad classification.

Engineering thermoplastics are generally divided into two classes, crystalline and amorphous. It is not the intent here in making the distinction between crystalline and amorphous polymers to probe the depths of molecular structure. This information is available[26–28] for those who want it in the vast literature devoted to the subject. Instead, only the practical differences will be covered.

TABLE 1.2 Commercial Blends

Blend	Trade name	Supplier
ABS/PC	Cycoloy	Borg-Warner Chemicals
	Bayblend MC2500	Mobay
	Bayblend MD6500	Mobay
ABS/PVC	Abson 89129	Abtec
	Polyman 509	Schulman
	Kralastic FVJ	Uniroyal
	Cycovin KAB	Borg-Warner Chemicals
	Cycovin KAF	Borg-Warner Chemicals
PVC/acrylic	DKE 150	DuPont
	Kydex	Rohm and Haas
PPO/PS	Noryl	General Electric
ABS/polysulfone	Arylon	Uniroyal
SAN/polysulfone	Ucardel P-4174	Union Carbide
CPE/PVC	Hostalit	American Hoechst

Source: Ref. 23.

Advantages of crystalline polymers[29] include:

Resistance to organic solvents.
Resistance to dynamic fatigue.
Temperature range increased by glass reinforcement.
Retention of ductility on short-term heat aging.
Orientation gives high-strength fibers.

Advantages of amorphous polymers[29] include:

Transparent.
Low and uniform mold shrinkage.
Low and uniform coefficient of thermal expansion.
Minimal postmold shrinkage.
Properties have smaller dependence on temperature.

Crystalline polymers have an ordered molecular arrangement, with a sharp, identifiable melting point. Due to the ordered structure of the molecules, crystalline polymers reflect most incident light and are thus opaque. In addition, these polymers undergo a significant reduction in volume when they solidify, resulting in high, although predictable, shrinkage. They are normally resistant to many organic solvents and have good fatigue and wear-resistant properties.

Amorphous polymers, on the other hand, have random molecular arrangements and melt over a broad temperature range compared to the crystalline polymers. A major physical difference is that light is transmitted easily through the polymer, making it transparent. Shrinkage is much lower with amorphous polymers. On the other hand, they are, in general, more sensitive to the effects of contacts with solvents and have poorer fatigue and wear characteristics.

The engineering plastics covered here will include examples of both crystalline and amorphous polymers, as can be seen from Table 1.3. In some end

TABLE 1.3 Selected Thermoplastics Showing Examples of Crystalline and Amorphous Polymers

Crystalline polymers	Amorphous polymers
Polypropylene (PP)	Polystyrene (PS)
Acetal	Acrylic
Fluoropolymers	Polycarbonate (PC)
Nylons	Polyphenylene oxide (PPE-based resin)
Polyphenylene sulfide (PPS)	Polysulfone
Polyether ether ketone (PEEK)	Polyether sulfone (PES)
Thermoplastic polyesters (PBT, PET)	

Source: Ref. 29. (Reprinted with permission of the American Institute of Chemical Engineers, New York.)

uses, only one or the other can be used; but in many applications, the properties of the two overlap sufficiently so that either one can be used.

Engineering thermoplastics, either crystalline or amorphous, are used in thousands of mechanical applications today in nearly every industry. As experience with them continues to grow, the mechanical uses of the engineering plastics continue to proliferate.

Some of the important reasons for this rapid growth, with particular emphasis on their use in place of metals, are the following.

Corrosion Resistance

Corrosion resistance was early recognized as a significant advantage of plastics over metals. An outstanding illustration of this advantage in action is the ball cock assembly used for controlling the water flow in toilets. The copper and brass ball cock assembly used in the past ultimately failed due to corrosion, the timing depending on the quality of the metal and the properties of the water. Today, many ball cocks are made of an acetal resin, and over 200 million of these have been used throughout the world with almost no failures due to corrosion.

Light Weight

At less than half the specific gravity of aluminum for most engineering plastics, the advantage of lighter weight is attractive for many applications in the automotive, airplane, appliance, and sporting goods industries. The automotive industry in particular has increased its plastics use from 15 lb./car to 200 lb./car in the last 20 years, with lighter-weight automobiles being one of the major objectives, although this includes both general-purpose and engineering plastics.

Low Cost

Over the years, as volumes have increased dramatically, engineering plastics costs have been reduced relative to the costs of such metals as magnesium, aluminum, and brass. Today, on a price-per-cubic-inch basis, a number of the plastics actually cost less than those metals with which they are frequently in competition. Further, because of the degrees of freedom provided by the injection-molding process, complex parts in plastics can be made in a single operation, in contrast to comparable metal parts that have to be assembled by joining a number of tediously formed metal parts. Of course, the reduction in labor costs obtained is significant. The increased life expectancy and greater ease of maintenance of the corrosion-resistant plastic parts are additional factors for the cost competitiveness over their metallic parts, even when the price of the plastic is at times higher.

High Strength-to-Stiffness Ratio

In the early days of engineering plastics, the replacement of metal parts to achieve lower costs and design flexibility was a viable objective, although the ratio of tensile strength to stiffness was not comparable to the die-cast metals the plastics were replacing. The replacements were made, however, when design calculations indicated that the properties of metals were clearly more than the application required.

More recently, the successful reinforcement with fiber glass or minerals of the mature and the new engineering plastics have dramatically increased the strength-to-stiffness ratio to one approaching that of the die-cast metals. Although the plastics still do not match the metals, their properties are much closer to metals than they used to be, resulting in a major increase in the range of applications in which the engineering plastics can supplant the metals.

Design Flexibility

The injection-molding process makes it possible to fabricate intricate shapes out of plastics, a distinct improvement in the versatility provided by metal-forming operations. In a number of large-volume applications such as automotive parts, the cost effectiveness of a material is governed not merely by the material cost but also by the improved processibility and design flexibility offered by the material. Use of polymers allows production of parts with complex shapes in a single molding operation, reducing the number of parts in design of a product and eliminating several assembly steps compared to a product made of metal components. In saving of installation cost, there are numerous examples of snap fits, molded-in hinges, and self-tapping screws. A specific example is the use of polycarbonate for a railway signal light made of nine molded components, replacing an assembly of 40 metal and glass parts, besides reducing the weight from 7 kg to 1 kg.[30]

Colorability

By incorporating dyes and pigments directly into the resin, it is possible to produce simply and economically items in an unlimited range of colors. This is another example of the versatility of plastics.

Coloring of plastic parts is important not only from an esthetic viewpoint but for other reasons as well. For example, it is well known that the surface temperature of any material exposed to sunlight depends on its color. Thus, if the plastic product is for exterior use, selection of color may be critical. In addition to imparting color, pigments offer significant protection against the damaging effects of sunlight. The wavelengths of light over the range 290 to 390 nm, which is in the ultraviolet (UV) region, are most destructive to plastics and promote polymer degradation by direct rupture of chemical bonds or by energy transfer sensitized by UV-excited impurities. Many pigments absorb UV radiation and act as screening agents. Others reflect and scatter light waves. Thus, the pigments help in preserving a part's physical integrity by substantially reducing cracking and loss of tensile strength.

Other valuable attributes of engineering plastics that are not related to metal replacement are the following:

Electrical Insulators

The dielectric properties of engineering plastics are outstanding and are responsible for the wide use of the plastics in the electrical/electronics industry.

Good Thermal Properties

Thermal conductivity of plastics is very low compared with metals. In appliances, for example, the thermal insulation provided for handles of cooking implements is an attractive, useful feature of a stiff, strong, high-heat, distortion-resistant, engineering plastic.

Transparency

Transparency is a feature now provided by several of the amorphous engineering plastics that makes new end uses possible. Example are the instruments and vessels used in medical and other scientific laboratories.[31]

THE NEED FOR A SELECTION PROCESS

Given the fact that engineering plastics are essential to the production of critical parts in many industries, how is the right plastic selected for a specific part?

Prior to the 1960s, the selection of a thermoplastic for an engineering application was simple, since the number of appropriate engineering plastics was limited. Even guesswork based on trial and error was possible. Today, there are literally thousands of grades of polymers and polymer modifications to choose from, and hence trial and error can be an extremely costly venture, if not an impossible mission. Polymer modification by way of blending, copolymerization, and through the use of fillers/additives has virtually increased the number of base polymer grades from hundreds to thousands, each grade with different properties and each suitable for different applications. Table 1.4 presents a picture of the variety of resins available these days, based on a very conservative estimate. This overwhelming variety of grades is at the same time a positive and a negative consideration in resin selection. While this proliferation makes

TABLE 1.4 Estimated Number of Available Grades for Selected Engineering Thermoplastics

Engineering thermoplastic	Number of grades[a]
ABS	482
ABS alloys	123
Acetal	245
Acrylic	64
Fluoropolymers	193
Nylon 6/6	936
Nylon 6	626
Nylon 11	48
Nylon 12	102
Nylon 6/10	41
Nylon 6/12	95
Polyamide-imide	8
Polyarylate	6
Polycarbonate	700
Polycarbonate alloys	38
Polybutylene terephthalate	248
Polybutylene terephthalate alloys	19
Polyethylene terephthalate	109
Polyethylene terephthalate alloys	13
Polyether ether ketone	41
Polyether-imide	82
Polyether-imide alloys	3
Polyether sulfone	52
Polyphenylene oxide-based resins	135
Polyphenylene sulfide	144
Polypropylene—filled	693
Polysulfone	65
	5311

[a]Authors' estimates based on *Materials Buyers' Guide of Plastics Technology.*

it simpler to find a fit for almost all reasonable applications, it does require more effort to discriminate.

A perfect example of the proliferation of candidate materials is the nylon family. The theoretical number of polyamide resins possible is quite high based on the condensation reaction of dibasic acids with diamines or the polymerization of lactams; and quite a few have been exploited commercially. Within this family are nylons 6/6, 6, 11, 12, 6/9, 6/10, and 6/12, each of which exhibits different properties and are used in different applications.[32] Tables 1.5–1.7 show

TABLE 1.5 Melting Points of Polyamides

Polyamide	Melting point, °C
Nylon 6/6	265
Nylon 6	226
Nylon 6/9	226
Nylon 11	184
Nylon 12	179

Source: Ref. 32. (Reprinted with permission of American Institute of Chemical Engineers, New York.)

the differences in the melting points, moisture absorption, and mechanical properties of some of these nylon types. Besides these, there are numerous copolymers of these basic polyamides. Furthermore, most of them are available in glass-reinforced, flame-retardant, and mineral-filled versions. These modifications can bring about a considerable change in the properties, as can be seen from Table 1.8. To this must be added nylons blended or grafted with toughening agents, such as elastomers and polyesters, and again there are filled- and glass-reinforced versions of these nylons.

While the nylon polymers were among the first of the engineering plastics to breed such a large family of available products, the newer[33] ones have quickly followed suit. The polycarbonates, polyesters, acetals, and other families also include a wide variety of molecular weights, molecular weight distributions, fillers, additives, reinforcing compounds, antifriction agents, antioxidants, UV stabilizers, toughening agents, etc., to present a bewildering multiplicity of choices to those charged with the responsibility of selecting the "right stuff."

It seems obvious that there should be a reasonable procedure available that will make it possible for those responsible for selecting the appropriate plastic to follow a logical process of discrimination. Within certain limits, this book has been prepared to suggest a screening process that is based on logic as

TABLE 1.6 Moisture Absorption of Polyamides

Polyamide	H_2O absorption, 24 h (ASTM D-570)	Percent amide
Nylon 6/6	1.0–1.3	38
Nylon 6	1.3–1.9	38
Nylon 6/9	0.5	32
Nylon 6/12	0.4	28
Nylon 12	0.25–0.3	22

Source: Ref. 32. (Reprinted with permission of American Institute of Chemical Engineers, New York.)

TABLE 1.7 Mechanical Properties of Polyamides

	Value of property				
Dry, as-molded property	Nylon 6/6	Nylon 6	Nylon 6/9	Nylon 6/12	Nylon 12
Tensile strength, kpsi	12.0	11.8	10.2	8.8	8.5
Break elongation, %	73	200	50	150	300
Flexural modulus, kpsi	400	395	335	290	165
Izod imp., notched; ft·lb/in.	1.0	1.1	0.7	1.0	2.0–5.5
Hardness, Rockwell R	120	120	61	114	105
Deflection temp. at 264 psi, °C	80	80	75	80	55
66 psi, °C	230	188	170	165	145

Source: Ref. 32. (Reprinted with permission of American Institute of Chemical Engineers, New York.)

well as experience, but with the caveat that there are certain empirical considerations involved that do not always reflect strict scientific principles.

A review of a resin screening process should include an understanding of the diverse backgrounds of those who originate the need for the plastic part. Consider that included in the list of final end users of the parts are such heterogenous backgrounds as investors, engineers, large corporations, entrepeneurs, scientists, toy manufacturers, mechanics, dress designers, boat designers, electronics specialists, and hospitals. The list of potential applications is nearly endless, and the degree of sophistication and the specific knowledge of the limitation of plastics vary widely. To begin with, the concept of a new object to be made of an engineering thermoplastic, or at least to have an engineering component, presents a challenge to some end users. Their approach to determining which plastic to use for their application will vary considerably,

TABLE 1.8 Key Properties of Reinforced Nylon 6/6

	Value of property with indicated reinforcement			
Property	None	40% kaolin	50% mica[a]	43% glass[a]
Tensile strength, kpsi	12.0	14.0	13.8	30.0
Break elongation, %	80	9		2
Flexural modulus, kpsi	400	870	1,900	1,600
DTUL at 1.8 MPa, °C	90	193	229	252
Reinforcement vol. %	0	23	29	26

[a]Properties measured in direction of orientation of reinforcement.
Source: Ref. 32. (Reprinted with permission of American Institute of Chemical Engineers, New York.)

from trying to follow the entire selection process from beginning to end personally, to the hiring of plastics design engineers.

For those without any experience with plastics, it is advisable to read the book, *What Every Engineer Should Know about Developing Plastics Products*[34] or seek out someone who can provide the kind of objective technical assistance that is required. One simple, and inexpensive, way to locate a reputable source of this kind of assistance is to contact any of the major resin manufacturers. These companies have access to the names of responsible plastic parts designers, and will usually be happy to provide their names and locations. At the same time, in some cases, they will be able to comment on the practicality of a proposed use themselves, although this might be only in general terms. For example, if the proposed use involves exposure to a continuous 1000°F environment, the material supplier could point out politely that the proposed application is simply not in the cards. On the other hand, if the application sounds reasonable, material suppliers can shed some light on what the next step should be and who should be contacted next to get the selection process moving, or they may volunteer to become further involved themselves.

Another major route to resin selection is through an experienced injection molder. Perhaps this route is the one most often used and, as a result, it will be covered in some detail.

Although there are some end users who have never been involved with engineering plastics before, most have had previous experience. Certainly the automotive companies have much experience and expertise, and have their own staffs and procedures for resin selection. Other large corporations in the appliance and electronics industries have similar advantages. Nevertheless, even these sophisticated industries occasionally have some problems selecting the exactly appropriate resin or, rarely, make a selection and later find it to be inadequate.

Finally, numerous small independent companies have a real need for resin selection assistance and must find it through material suppliers, injection molders, or designers who have the necessary knowledge and experience.

The successful production of a defect-free, high-performance, functionally useful plastic article requires the proper interplay among the following four variables: (1) part, geometry, (2) resin, (3) processing characteristics, and (4) mold design. Hence the process of resin selection requires teamwork among four groups: (1) the product designer and supplier, (2) the material supplier, (3) the processor, and (4) the mold maker. There is an obvious interdependence among the members of these groups, and the lines of communication must be kept open and well defined at all times. Poor communication can result in inferior-quality articles.

Many times the processor fails to put the material supplier/specialist in contact with the product designer/supplier or the mold maker. This results in incorrect or incomplete information being transferred to the material selector during his evaluation and recommendation. In reality, all members of the team must interact and provide detailed information, which ought to be carefully studied before recommending the right thermoplastic for a particular application.

Those experienced in the design and use of parts molded from engineering plastics are certain to have serious reservations about the practicality of trying to define a selection process. They well know that part design, molding cycles and conditions, the amount of recycled material in a part, etc., all can have a major effect on the ultimate utility of a plastic part. On the other end of the spectrum are those who are considering the use of an engineering plastic for the first time, and who have no idea what approach to use to select a resin.

It is the intent of this book to provide guidance in such a way that both those experienced in the use of engineering plastics and the novice with a complete lack of previous experience can at least appreciate the considerations involved in resin selection and become active in the decision-making process.

2

Beginning the Selection Process

INFORMATION NEEDED TO BEGIN THE PROCESS

The practical starting point of the resin selection process begins with a series of questions about the proposed new plastic application that should be evaluated by the resin selector.

General Information

What is the function of the part?
How does the assembly operate?
What is the part geometry and configuration?
Are there space or weight limitations?
What is the required service life?
Can several functions be combined in a single part?

Can the assembly be simplified?
What are the consequences of part failure?
What is the quantity and speed of production of the part?
Are there any special processing requirements or constraints?

Codes and Specifications

Are acceptance codes (e.g., Society of Automotive Engineers, Underwriters Laboratories) required?

Costs

What are cost and pricing limitations on the part?

Environmental Considerations

Chemical environment
Exposure to sunlight and weathering
Humidity
Operating temperature

Mechanical Considerations

How is the part stressed in service?
What is the magnitude of the stress?
What is the stress-versus-time relationship?
What is the maximum deformation that can be tolerated?
What are the effects of friction and wear?
What tolerances are required?

Electrical Considerations

Voltage requirements
Tracking requirements
Insulation requirements

Appearance Considerations

Style
Shape
Color
Surface finish

These are basic questions, and answers to them are to be provided by different members of the team, namely, the product designer/supplier, material supplier, processor, and mold maker, by way of information as listed in Tables 2.1–2.4. This pool of information governs the remainder of the procedure required to reduce the number of possible resin candidates to a manageable level.

The next step is to begin to provide detailed responses to some of the questions posed in the information-generating phase. This can be done by considering in depth the following factors, in the order they are presented.

TABLE 2.1 Information to Be Provided by Product Designer/Supplier

Function of the part
Minimum properties necessary
Desirable features
Part geometry and configuration
Cost parameters
Ambient operating conditions
Surface appearance
Production quantity and speed
Specific requirements
Tolerances
Testing and prototyping
Assembly operations

TABLE 2.2 Information to be Provided by Material Supplier

Resin class and type
Properties and limitations
Necessary design changes
Design calculations
Resin behavior and properties
Under ideal testing conditions
Under required operating conditions
Processing requirements
Optimum mold design
Resin costs (cents/in.^{3})
Additional testing
Tolerances

TABLE 2.3 Information to be Provided by Processor

Manufacturing techniques
Unit cost of finished part
Production schedule
Quality control and testing
Practical tolerances
Mold maintenance

TABLE 2.4 Information to Be Provided by Mold Maker

Mold design
Mold technology
Tolerances and dimensions
Mold costs
Delivery schedule
Mold quality
Mold testing
Necessary changes

MATERIAL SELECTION FACTORS

A number of factors are involved in selecting the appropriate plastic for a proposed use, some scientific and logical, some legal, and some pragmatic and not so logical. It would be much simpler if only science were involved, of course, but realistically that is not the case. Where science is involved, the following factors are the major considerations in the selection process:

1. Codes and Specifications
   Agencies that regulate safety:
   a. Food and Drug Administration (FDA)
   b. United States Drug Administration (USDA)
   c. National Sanitation Foundation (NSF)
   d. Underwriters Laboratories (UL)
   e. National Association of Homebuilders (NAHB)
   f. Federal Aviation Administration (FAA)
   g. Federal Motor Vehicle Safety Standards (FMVSS)
   h. Plastics Pipe Institute (PPI)

Specifications:

a. Military
b. Federal, state, and local
c. Industrial (automotive, electrical)

Code bodies:

a. American Society for Testing and Materials (ASTM)
b. American National Standards Institute (ANSI)
c. Society of Automotive Engineers (SAE)

2. Costs
   a. Total part cost
   b. Value-in-use
3. Service environment
   a. Chemicals/solvents
   b. Outdoor exposure
   c. Water/humidity
   d. Temperature
4. Structural properties
   a. Strength
   b. Stiffness
   c. Toughness
   d. Creep
   e. Fatigue
5. Other design considerations
   a. Friction and wear
   b. Electrical properties
   c. Dimensions
   d. Assembly
   e. Esthetics
   f. Consistency of properties
6. Process considerations
   a. Right process
   b. Right polymer grade for the process

Codes and Specifications

Often codes or formal specifications for an article have been prepared by military or regulatory agencies, resulting in de-facto go/no-go determinations. The end user or designer must be certain that the part to be produced and marketed either does not require any of these specifications or, if it does, exactly what the specification is. In selecting a candidate resin, one should first contact material suppliers and ask the direct question: "Do you have any products that meet the stated specification?" Although it is sometimes possible to find this out in the process of researching the specification, the safest and most direct route is to ask material suppliers. As a matter of fact, many times it is very difficult to interpret some specifications, while material suppliers normally have carefully researched them to determine which of their products apply. If material suppliers have more than one product meeting the specification requirements, the end user should ask which of the candidates is the most appropriate for the particular application.

Costs

In this, the very first phase of the resin screening or selecting process, not enough information will be available on potential resin candidates to make detailed cost analyses on the final product. Nevertheless, go/no-go values must be estimated for the maximum allowable cost of the proposed part, as well as the most realistic selling price of that part.

This is obviously basic. After developing the selling price that will make the venture viable and attractive in the marketplace, it is clear that some value below that must be allotted to the cost of the resin. For example, if the part will sell for no more than, in effect, $3/lb, it is clear that all resins costing more than $3/lb must quickly be eliminated from consideration. If the part is not to be marketed per se, but rather as an integral part of a more complex and expensive product, then the go/no-go elimination approach does not apply; and it remains for more quantitative part cost determinations before a decision can be made on which candidates to eliminate from consideration.

In any case, part cost must be carefully estimated *after* the potential resins have been reduced to a reasonable number. A technique to simplify this calculation will be suggested later.

Value-in-Use

Actually, the situation in selecting a resin will normally result in concentrating the choice into a few that seem to offer comparable environmental properties and similar costs. It is then that a more complete examination of the value-in-use properties must be made.

The economic incentive for using one resin over the others being considered can be the difference between the installed cost in that resin and, perhaps, an existing assembly, or one made of the other resins reviewed. As an example, if one resin never needs lubrication in use and the others need an expensive lubricating operation frequently, it is possible that, although the one not requiring lubrication is slightly more expensive than the other candidates, its value-in-use is better. For final determination of this, a reasonable estimated cost of the final part is necessary, and included in a later chapter is a suggested simple cost estimating procedure.

Service Environment

An important step in the process of selecting the right engineering thermoplastic for an application is the distillation of the available resins into a heart cut based on properties that are not subject to manipulation by design. However, the ability of a given plastic to retain adequate properties in the presence of chemicals or solvent agents is not a simple thing to research in available literature. Obtaining reliable chemical resistance data for plastics and using them for material selection criteria is more complicated than a similar selection process for metals. There are several reasons for this.

Chemicals/Solvents

Each family of polymers varies greatly from the others in the number and type of chemicals to which they are resistant. Further, even within a family, certain individual members may differ. Therefore, while certain reagents such as mineral acids or aromatic hydrocarbons will attack most members of a particular family, it is not safe to generalize to every member of that family. It is also often erroneous to assume chemical attack based only on chemical similarity of reagents. The chemical resistance data in the *Modern Plastics Encyclopedia*, 1978–79 ed., pp. 500–560, are tabulated by individual trade name and grade of plastic, and by specific chemical identity and concentration of reagent. The 1979–80 *Modern Plastics Encyclopedia*, from which a large part of this is quoted or abstracted, includes a comprehensive chapter on the chemical resistance considerations of plastics, on pp. 489–494.[35]

Another major consideration of chemical resistance of plastics is that they interact with chemical environments that vary in rate and extent of effect. The mechanisms of chemical attack include chemical reaction, solvation, absorption, plasticization, and stress cracking. The different mechanisms affect different performance properties and produce different degrees of risk that vary with the application.

It is true that, for economical screening of candidate resins, a review of published data summarizing the apparent resistance to various chemicals by the polymers is helpful. However, if a part is to be used in a questionable environment, and the literature is not sufficiently definitive, it is prudent to obtain molded test specimens of the candidate plastic and subject them to the specific environment involved.

Chemical Reaction

Chemical reaction involves the direct chemical reaction of the active agent on active sites along the polymer chain, or at the chain ends, which, depending on the severity of the reaction, can reduce physical properties, or even destroy the polymer completely. Usually, the extent of the attack will be indicated by a reduction of molecular weight. Chemical reaction on plastics is obviously accelerated by increasing temperatures, and any testing of parts should be carried out at temperatures anticipated for the part's use.

Solvation, Absorption, and Plasticization

Although engineering plastics are difficult to dissolve in most solvents, they can be affected by them, resulting in swelling and in weight and dimensional changes. Of course, the strength, stiffness, and creep resistance are also affected. In the long run, if solvation and plasticization appear to be an incipient problem, more specific environmental tests on the polymer are suggested.

Environmental Stress Cracking

Environmental stress cracking (ESC) is the name given to the phenomenon by which a plastic under high stresses may crack in contact with certain active environments such as detergents, fats, and silicone fluids. A number of chemical and physicochemical effects are involved in any given ESC phenomenon.[36] Plastics may be subjected to mild stress and strain without failure, or they may be subjected to weak solvents or aqueous surfactants without failure, but when they are subjected to a combination of these two types of variables they often undergo ESC and there is catastrophic brittle fracture.[37]

Environmental stress cracking is the most frequently encountered, and the most important chemical attack problem. The active agent causing the failure does not have to be a solvent for the polymer to cause failure when a part is placed under stress. The rate at which an agent will cause stress cracking varies considerably, and depends on the agent, its concentration, the plastic, the temperature, the stress level, and other factors such as surface roughness, sharp corners, and polymer morphology, which are not completely understood.

Although a review of available literature might suggest certain reagents that will cause stress cracking of individual polymers, such a list is by no means complete. In the early stage of resin selection, the existing list can be helpful in screening out probable gross failures, but end-use testing under stress is inevitably the only route to confidence that the right resin has been chosen.

Outdoor Exposure

Weatherability is a difficult resin selection criterion to rate, since all the major polymers now are provided in UV-stabilized compositions. However, some discussion of the subject is necessary, since there are so many outdoors applications for engineering plastics.

It should be helpful to describe what is meant by the term *weathering*. Weathering is a process that includes the effects of light, oxidation, and temperature. The natural events of rain, wind, humidity, and atmospheric "pollutants" such as ozone also have considerable effect. To simplify weathering, it may be said that it is an oxidation accelerated by the high energy level of ultraviolet light. Oxidative and nonoxidative photodegradation have deleterious effects on the chemical structure of plastics[38] as well as on the mechanical properties.[39]

Although only about 4% of the total radiation reaching the earth is in the ultraviolet, it is the ultraviolet energy that is primarily responsible for the changes that occur when plastics are exposed to the weather. The rate at which plastics degrade depend on a number of factors, all of which can and do vary with geographic location and season.

Fortunately, since nearly 20–25% of all plastics used are exposed to the outdoors, for more critical applications additives have been developed that inhibit UV degradation, some of them significantly.[40–42] Probably the best of these is carbon black, and if, as in the case of the classic statement attributed to Henry Ford, "You can have any color you want as long as it's black," you can use a black part, then a plastic containing well-dispersed carbon black will endure in the outdoors. If colors other than black are required, some common ultraviolet stabilizers can function by screening out light, by absorbing the ultraviolet, as deactivators, or as free-radical inhibitors. Pigments can be incorporated in such stabilized resins to impart the desired colors.

Normally, it is impractical to run actual weathering tests on candidate resins for an application due to time and expertise considerations. Instead, information should be sought from material suppliers and their recommendations accepted.

Water and Humidity

In many applications, engineering plastics are required to be immersed in water at all times. Most polymers can be useful under such conditions. However, when the end use calls for temperatures above normal ambient (usually considered to be 73°F), considerable discrimination must be used. For such conditions, particularly when the water temperatures is to exceed 125°F, some polymers should not be used. One example of such a polymer is polyethylene terephthalate (PET polyester), which is not greatly affected by immersion in water at normal temperatures, but at elevated temperatures hydrolyzes to such an extent that the physical property loss is extensive. Polycarbonate is another polymer that loses properties significantly when exposed to hot water for long periods of time.

For those engineering plastics specifically reviewed later in this book, there will be a statement referring to limitations (if any) of the individual resins in contact with water at elevated temperatures. For other plastics *not* covered here, the resin selector should contact the appropriate material supplier when water contact is a consideration.

In contrast to water immersion, humidity is not a serious problem with regard to loss of properties, although several engineering plastics are somewhat adversely affected, especially under warm humid conditions. The warm humidity resistance of more than a dozen engineering thermoplastics has been evaluated,[43] and the database provided can serve as a basis for selecting plastics for use in warm, humid environments.

A more commonly observed effect is water absorption from the atmosphere by some polymers. The most common of these are nylons, which gain or lose moisture according to humidity changes in the atmosphere. Of course, such changes with changes in humidity are not rapid, and do not respond significantly to daily humidity variations. Nevertheless, at the average humidity of 50%, the nylons absorb 2.5–3.0% water. When this happens, there usually are dimensional changes in a molded part, and there will also be major changes in such structural properties as stiffness and strength.

It is important, then, in the process of resin selection to check the effect, if any, that water and/or humidity may have on candidate resins.

Temperature

The end-use temperature of a proposed new application is frequently one of the most important criteria for the polymer to be selected. Temperature can affect the properties of the spectrum of polymers in different ways. For example, some of the major properties affected are modulus (stiffness) and tensile strength, which decrease significantly with increasing temperatures, while elongation and impact properties improve. The resin selector and/or designer must design parts for the maximum expected use temperature and, of course, must select a polymer that will have adequate properties at that temperature.

In connection with general temperature considerations, it should be kept in mind that most engineering plastics, particularly nonreinforced versions, have high coefficients of expansion, and if a proposed application will involve the use of metals and the plastic as part of a single item, care in design is required to allow for the difference in expansion between the two as temperature changes.

The need to hold plastic parts at elevated temperatures for extended periods of time is a frequent occurrence. However, heat aging can cause eventual deterioration of a polymer's properties, or even complete failure of its structure. Obviously, these possibilities must be considered carefully in the resin selection process, and careful attention paid to the product literature available on the various resins.

Some properties of engineering plastics will, of course, be affected by below-normal (normal being considered to be 73°F) temperatures as well as elevated temperatures. When parts are to be designed for a cold atmosphere, the product literature will normally provide sufficient guidance on the effects of low temperatures on the properties of the candidate resins. It is important to pay particular attention to impact resistance as temperatures drop below freezing, because brittleness is not easily predictable. The toughness of some resins is drastically reduced by a drop of a few degrees in temperature in the 32°F range, while others are only slightly affected until severe cold is encountered.

Attention should also be given to the combined effect of vibrations and end-use temperatures on the mechanical properties of a polymer in any given application. Such information is made available through dynamic mechanical data[44–46] which provide insight into the response of the material to forced oscillations and its capacity to dampen vibrations at different temperatures.

Structural Properties

A number of important structural properties of plastics must be considered by designers as they pursue the resin selection process, and later as they proceed with the more definitive step of finalizing part design. However, there is no intent here to become involved in the mechanics of part design per se. Instead, simplistic definitions of various structural properties are presented so that their functions can be understood by decision makers who are not skilled in design, and so that a comparison of polymers can be made based on property values. (In Chapter 3, the basic procedures for determining these structural property values are presented to provide a more complete understanding of their significance.)

From a practical point of view, it is preferable to run definitive property tests in a short time. Actually, the important standard tests involving mechanical loading can usually be run in about 15 min, and these are said to define "short-time" properties. On the other hand, designers must be able to predict how a plastic will perform after long periods of time, and to do this they need some measurements of "long-term" properties; others connected with the decision-making process must also have some feel for what they are. A method has been suggested[47] for determining the long-term mechanical behavior of plastics through short-term tests. It has been shown how curves of stress versus log time creep can be combined into master curves which are independent of percentage strain and the temperature at which the creep data were obtained. The advantage of the master curve is in predicting the creep behavior of a particular specimen for a year through generation of creep data for only a day. The suggested approach[47] has immense utility in design calculations and a high degree of economic benefit.

Short-Term Structural Properties

Tensile strength is the amount of force (or pull), in pounds per square inch, required to elongate the plastic by a defined amount. It is basically the resistance of a material to being pulled apart. Thus, the higher the value, the stronger is the material. To pull plastics apart may require anywhere from 1000 to 50,000 psi; for metals the values can be as high as hundreds of thousands of psi.

Compressive strength provides a measure of the maximum load in pounds that a 1-in. section of the material will support without fracturing.

Modulus is the number of pounds per square inch required to cause deformation in terms of elongation, compression, bending, or twisting. It thus represents the stiffness of a material subjected to tensile, compressive, flexural, or torsional action. For example, flexural modulus is a measure of the stiffness of a plastic during the first part of its bending process. The higher the reported value, the stiffer is the material.

Heat deflection temperature is the temperature at which a plastic sample bends a specific amount when a given load is applied. Usually the load is specified to be either 66 or 264 psi. The higher the reported value, the higher the plastic's practical end-use temperature is likely to be. At present, the ASTM heat distortion test can provide more than just a single-point measurement. It has been shown[48] that the single-point heat deflection temperature value obtained under ASTM D-648 can be used as a normalizing factor to create heat deflection temperature-versus-stress curves for each generic type of polymer. The advantage of such curves lies in their ability to provide a quick estimate of the behavior of the material at other stress levels useful for design purposes.

Izod (notched) is a relative measure of a plastic's resistance to impact. It consists of breaking a sample under controlled conditions, and reporting the results in terms of foot pounds per inch required to break. The higher the reported value, the tougher is the material. Although many engineers feel that this test is inadequate, it is still frequently used and referred to.

Elongation at break is the amount, reported in percent, that a plastic stretches or elongates before it breaks during controlled force loading test conditions. It can also be construed as an indicator of toughness. The higher the reported value, the more the material elongates. A material with a higher tensile strength does not necessarily indicate a tougher material. In fact, a material that has high tensile strength and relatively high elongation is tougher than one having high tensile strength and low elongation.

Long-Term Structural Properties

Creep[47] (or deformation under load) is the deformation of a plastic part that takes place over a specified long period of time under controlled conditions while the part is supporting a load, and is reported in terms of percent deformation. The lower the reported value, the greater resistance the material has to deformation.

Fatigue endurance is a measure of the ability of a plastic material to resist failure (breakage) after being subjected to repeated loads at different levels of loading. The higher the reported value, the more resistant the material is to conditions that lead to fatigue failure.

Other Design Considerations

Design is the first stage of manufacturing. It is essential to give attention to a number of other design considerations which have an affect on material selection.

Friction and Wear

In making material selection, a good deal of thought ought to be given to the friction and wear that the processing equipment is likely to experience in day-to-day operation.[49] Especially when using fillers which are fibrous or acicular in the plastic, an abrasive environment is created which can cause considerable machine wear with time.[50] Often, the thermal decomposition of certain chemical additives in the plastics can cause corrosive damage. The wear due to fillers is related directly to filler hardness, particle size and shape, extent of agglomeration, and level of loading. It is often beneficial to keep the filler loading to the minimum level possible without sacrificing product properties. However, if high loading is an essential requirement,[22] then the suggested method to reduce wear is to make use of hardened nitrided steel or to put a special borofuse coating on the processing equipment parts that come in contact with the abrasive fillers. If this aspect is not given due consideration during equipment design, recurring capital expense and deteriorating product quality with time will be a natural consequence.

Electrical Properties

For any plastic product which comes in contact directly or indirectly with an electrical current, the electrical properties of the plastic must be considered. When using the plastic as an insulator, the following properties must be evaluated.

Dielectric strength is a measure of the electrical breakdown resistance of a polymer under an applied stress and depends to a measurable extent on the temperature, preconditioning, and thickness of the material.

Volume resistivity, expressed in ohm-centimeters, is a measure of the resistivity of electrical dc current through the thickness of a specimen. Materials with resistivities above 10 Ω·cm are considered insulators.

Dielectric constant is the ratio of parallel capacitance of the material to that of an equivalent volume of vacuum.

Arc resistance is the time in seconds that an arc may play across the surface of a material without rendering it conductive.

Dimensions

It is important to be able to mold a plastic part to correct dimensions within tolerance limits. For this, it is necessary to have knowledge of the mold shrinkage and the coefficient of thermal expansion of the plastic.

Mold shrinkage of the plastic material must be accounted for during the design of the mold. The material supplier normally provides information on mold shrinkage, but these are only average or near-about values. Shrinkage of a plastic in the mold may be different for material flowing lengthwise with the gate than for material flowing crosswise to the gate. If fibers are present, shrinkage is reduced; and if fibers orient during molding in a preferential direction, then shrinkage in that direction will be different. Hence, careful consideration to mold shrinkage is important.

The coefficient of linear thermal expansion indicates the amount by which the material will grow when heated. It is normally expressed in inches per degree. The thermal expansion of plastics is about eight times greater than that of other engineering materials. Hence, when plastics are used in conjunction with metal parts, the differences in their coefficients of linear thermal expansion must be borne in mind when making design decisions. If expansion of the part is restricted in any way, such as being attached to a framework of a material that has a lower thermal expansion coefficient, then stresses will develop and result in warpage or distortion as a means to relieve the stresses.

Assembly

Proper consideration of assembly at the design stage[51] is known to result in 20–40% lower manufacturing costs. Early selection of the assembly is important because a manual assembly differs very much from an automated assembly in the requirements that it imposes on product design. The cost of assembling a product is related to both the design of the product and to the assembly process used for its production. When the product design is such that it can be assembled by the most optimum process, then the assembly cost is the lowest.

Esthetics

Many times, the injection-molded finished part shows knit or weld lines, sink marks, and surface roughnesses which are prominent enough to spoil the esthetics of the product. In such cases, a coat of urethane paint is often applied to cover the molding imperfections. Of course, this secondary operation is expensive and time consuming; hence the preference is to take care to produce defect-free products by proper design, optimum processing, and use of expert systems for troubleshooting.[52,53] The causes of molding defects are related to raw materials, molding operations, and the mold, and hence these must be taken into consideration.

Consistency of Properties

It is very important that the manufactured plastic product shows consistency in the properties that are relevant. In other words, quality control is an important consideration. The reproducibility of product properties depends largely on the consistency of the quality of raw materials, proper control of the processing operation, and the condition of the hardware.

Process Considerations

Sufficient attention must be given to the process and the manner in which different materials process. It is important that the right choice of process is made based on the product needs and that the right choice of the polymer grade is made to ensure that it suits the process and the processing conditions.

Right Process

Nearly all thermoplastics can be processed on any of the available thermoplastic processing equipment. However, the idea is to make the best choice based on the needs of the product design. At times, the product manufacturer already has a particular type of processing equipment in use and hence wishes to use the same equipment to produce a new product, in order to save capital expenditure. In such cases the process is predetermined and the selection of the plastic has to be done based on these existing equipment considerations.

On the other hand, sometimes the flexibility of choosing the process may be available. In such cases it is important to investigate all the parameters, including tooling costs, tolerance requirements, raw material availability and part costs, before settling on a particular process. When a review of all processes that seem to have merit is done, one or possibly two methods are likely to emerge as appropriate. These then may be thoroughly evaluated and the final right choice made.

Right Polymer Grade for the Process

The grade of the polymer is an important consideration when making the final material selection. Only particular grades of each polymer flow appropriately in particular processing equipment and can be transformed into defect-free products. The rheological properties of the polymer under the conditions of stress and temperature that it experiences in the processing equipment need to be thoroughly understood. The melt flow index (MFI) of the polymer gives some idea of its flow behavior during processing, and this value is readily available from material suppliers. Of course, the desirable information is the entire viscosity-versus-shear rate curve at various temperatures so that flow behavior under all conditions experienced by the polymer in the processing equipment can be clearly understood. A method for generating the entire rheogram from MFI is available, and this can be used effectively for estimating processing parameters during polymer flow through the processing equipment.[54]

3

Physical Property Data

INTRODUCTION

In subsequent chapters, considerable emphasis will be placed on the consideration of structural and other properties of engineering plastics, including an attempt to rank a number of plastics according to these properties. This is basic to the selection of the appropriate resin.

Many design engineers and other technical people involved in the selection process are accustomed to using published property data, but it is unlikely that many of them have devoted much thought to how the data are obtained, and in some cases, how to interpret them. Further, end users and others not that familiar with plastics property data and their interpretation should at least understand the fundamentals of them. For these reasons, this chapter is devoted to short descriptions of some of the major physical property tests and their interpretation.

One consideration of critical importance in resin selection is the understanding of the difference between the elasticity of metals and the viscoelastic nature of engineering plastics. Although such design criteria as tensile strength and elongation, for example, are critical to the proper design of a functional part, the designer in plastics must use them with the understanding that such data provided by material suppliers in good faith are not absolute. The "constants" supplied in the literature are not constants in the true sense, since they vary with time under load, rate of loading, temperature, etc. Furthermore, even between material suppliers there is not always agreement on the details of the testing methods, in spite of much effort on the part of testing societies to establish precise methods.

Material suppliers occasionally tend to follow the specific requests of designers for the type of property information needed on the polymers, rather than initiate new specifications themselves. Only when they find they have a new resin with certain specific, uniquely advantageous properties will they add to the already burdensome list of physical property data.

The designer and end user must understand that it is expensive to establish a meaningful new property criterion. Usually, years are required to develop such a test, since possible effects of time on the property must be examined, to say nothing of defining absolutely the environmental control required for reproducible test results.

In addition to the tests themselves, major consideration must be given to the preparation of test specimens. The injection-molding process is normally used to prepare test specimens, and injection-molding variables that can affect the values obtained in the tests are many. Sometimes control of all of these variables within a single laboratory is uncertain. It is difficult to obtain precisely reproducible results between different molding machines within a single laboratory and between laboratories within one supplier's several locations, to say nothing of between competitive suppliers.

Melt temperature, hold-up time, injection speed and pressures, gate design and location, mold temperature, specimen thickness, and the anisotropic nature of some polymers are just some of the variables that can affect a specimen, even before it is subjected to the variables of the testing procedure. Efforts are made to define and adhere to the molding variables within a supplier's laboratories, but not all of them are specified in independent testing procedures. Certainly there are going to be some differences between suppliers in their specimen preparation efforts.

Beyond that, if the test is accepted as a necessary, or even routine, indicator of a design criterion for a plastic, the material supplier is faced with the quality control aspect of production. How often must the test be run in-house, should it be based on a routine continuous sampling basis, a blend perhaps, or a "lot" basis? Such testing can add thousands of dollars per year to the cost of producing a product. With this in mind, it becomes understandable that material suppliers are not anxious to develop and provide data on a new physical test on a routine basis, unless it helps significantly in improving the dependability of designers' decisions in their work.

In spite of all these difficulties, specification data must obviously be provided by suppliers. When specification are available, it can be assumed that the data represent the average of a vast number of tests performed, and that the supplier is comfortable with them.

A GUIDE TO STANDARD PHYSICAL TESTS FOR PLASTICS

The tests selected here for description are some of those commonly listed in technical literature. No attempt has been made to cover all ASTM tests. Complete information is given in the book, *ASTM Standards on Plastics*, which is available from the American Society for Testing and Materials, 1916 Race Street, Philadelphia, PA.[55]

In addition to describing the tests, a number of terms are defined and, in certain cases, practical problems of carrying out the procedures are mentioned. The tests are described as carried out on standard test specimens. Often these procedures are modified for testing commercial parts. When that is the case, somewhat different results should be expected, and care will be required in interpreting the data.

It should be understood that many physical tests are normally subject to fairly large errors. As a rule of thumb, the error of testing should be considered ±5%. For this reason, judgment is required in evaluating small differences in test data to make certain a real difference exists. Many errors are associated with the test itself, while others arise through not having perfect or even representative test pieces.

Tensile Properties (ASTM D-638)

Tensile properties refer to the behavior of a material when it is subjected to forces which tend to pull it apart. By the proper choices of test conditions and recording sufficient data, it is possible to determine many of the tensile properties through a single test.

ASTM tensile properties are almost always determined on a standard dumbbell-shaped specimen. The most common specimen defined by D-638 has an overall length of 8½ in. and a center section 2¼ in. long, with constant cross-sectional dimensions of ¼ in. (See Figure 3.1.)[56]

In this center section, a 2-in. length is marked off with wax pencil marks, called *gauge marks*. It is actually this portion of the sample within the gauge

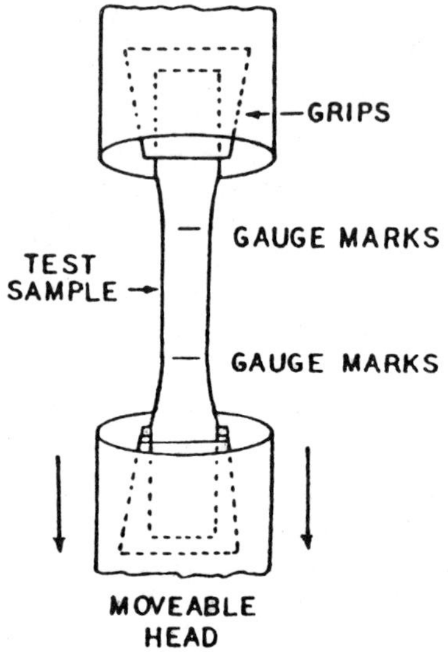

FIGURE 3.1 Sketch of tensile strength test assembly. (From Ref. 56.)

marks which is under test. Results from specimens failing outside the marks are discarded.

The specimens are tested in a machine giving a straight tensile pull without twisting or bending. The Instron* universal testing machine is frequently used for this purpose. One jaw is held stationary and the other is in a movable crosshead. The D-638 method specifies that the samples be pulled at a *constant* rate of crosshead movement (0.2, 2.0, or 20.0 in./min, depending on the type of plastic under test).

Two properties are measured while the sample is being pulled apart: tensile stress and tensile strain. *Tensile stress* is the strength of the pull within the area between the gauge marks, based on the original cross-sectional area. It is expressed in terms of pounds per square inch. *Tensile strain* is a measure of how much the sample has been stretched by the pull. Again, only the length between the gauge marks is considered. Tensile strain is calculated by dividing the extended length between the gauge marks by its original length. Thus, strain is a ratio (in./in.) and does not have dimensional units.

During the first part of the pulling process, both the tensile stress and the tensile strain continue to increase, and in proportion. When this takes place, the material acts like a spring, and is said to exhibit elastic behavior. The ratio of

*Instron® is a trademark of the Instron Corporation of Canton, MA.

tensile stress to tensile strain is called tensile modulus, to be discussed later. Some materials, such as methacrylates, will break when they have been strained (stretched, elongated) only a small amount, and while still showing essentially elastic behavior. Other materials, such as nylons and polycarbonates, can be stretched many times their original length before they break. The latter have what is called a *yield point*, and a corresponding *yield stress*.

The yield point is reached when the material continues to elongate (strain) with no increase in the strength of the pull (stress). When this takes place, the material yields, is permanently deformed by elongation. The tensile stress at which yield takes place is called *yield stress*.

The tensile behavior is perhaps most easily understood from a study of a stress/strain diagram, as illustrated in Figure 3.2.[57] The material represented by curve A does not have a yield point. Such a curve would be obtained when testing acrylic resins, such as Lucite®* or Plexiglas®.† The material represented by curve B does have a yield point. Nylon or acetal polymers would have such curves. The ends of each of the curves are the breaking points of the samples.

It is fairly common practice to report four results from a standard tensile test: *yield stress*, *tensile strength*, *ultimate strength*, and *percent ultimate elongation* or *elongation at break*. *Yield stress* has already been defined as the tensile stress at the yield point. It may also be called yield strength, but ASTM has a more specific definition for yield strength. *Tensile strength* is the maximum

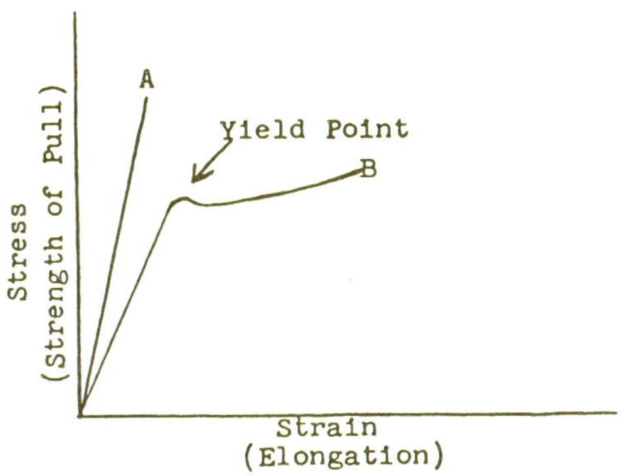

FIGURE 3.2 Diagram of stress/strain of sample under tension. (From Ref. 57.)

*Lucite® is a trademark of E. I. du Pont de Nemours & Co.

†Plexiglas® is a trademark of Rohm and Haas.

tensile stress which is observed while the sample is being pulled. It may or may not be the tensile stress at the time of failure, which is the *ultimate strength*. In some cases, the tensile strength is equal to the yield stress. The *elongation at break* is equal to the tensile strain at failure multiplied by 100.

As the rate of pulling the samples affects these results, standard rates of pulling should be used. If a nonstandard rate is used, the fact must be included in any references to the test results. For most resins, ASTM usually specifies what the speed of the crosshead must be.

Flexural Properties (ASTM D-790)

Flexural modulus and *flexural strength* are two important design parameters, and are determined as specified in ASTM Method D-790. The terms flexural modulus and flexural strength relate to stiffness, or resistance to bending, and to the force required to break a sample by bending or flexing. Flexural properties are determined on specimens supported near each end and loaded in the middle, as illustration[56] in Figure 3.3.

Although ASTM D-790 suggests several sizes of specimens with corresponding lengths between supports (span) that can be used, a ½ × ⅛-in. bar with

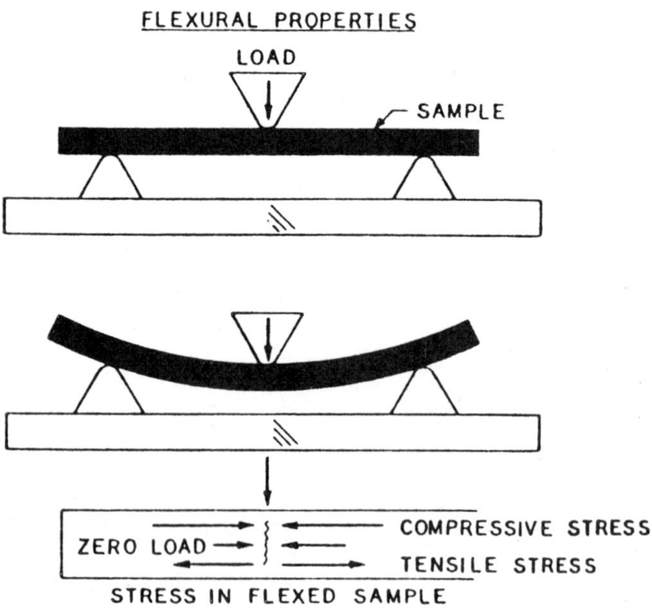

FIGURE 3.3 Sketch of flexural properties test assembly. (From Ref. 56.)

a 2-in. span is the size used most often. For these specimens, ASTM D-790 specifies that the rate of motion of the rod providing the load in the test should be 0.05 in./min. An Instron testing machine will provide this controlled motion nicely. As the test proceeds, the force of the load and the amount of deflection (bending) are measured (See Figure 3.4.)[57]

When the sample is bent in this test, forces of resistance (stress) are set up within the plastic which try to restore it to its original flat condition. Further, as a result of the bending, the material is distorted (strained). These

FIGURE 3.4 Photograph of test specimen in testing apparatus for determining flexural strength. (From Ref. 57.)

stresses and strains are more easily understood from an examination of a bent specimen as shown in Figure 3.5.

Before any load is applied to the sample, lines AB, CD, and EF are, of course, equal in length. When the load is applied, line AB becomes the shortest and EF the longest, while CD, which is an imaginary surface midway between the two lines, does not change in length. When the outside line EF does become longer with bending, its surface is stretched, and when the inside line AB becomes shortened, the surface is compressed.

Actually, there are increasing degrees of stretching from the midline toward the outside of the bend and increasing amounts of compression from the midline to the inside. The compression of the inside surface is equivalent to the extension of the outside, with respect to both stress and strain. Flexural properties are calculated and reported in terms of the maximum stress and strain that occurs in the outside surface.

The *maximum stress* (restoring force caused by bending) is related to the load and the sample dimensions by the following formula:

$$S = \frac{3p^*l}{2bd^2} \tag{3.1}$$

where

S = stress (psi)
p^* = load (lb)
l = length of span between supports (in.)
b = sample width (in.)
d = sample thickness (in.)

The corresponding *maximum strain* (stretching or compression is related to the amount of bending (deflection), and to the sample dimensions by the following formula:

$$\gamma = \frac{6Dd}{l} \tag{3.2}$$

where

γ = strain (in./in.)
l = length of span between supports (in.)
d = sample thickness (in.)
D = deflection (in.)

Flexural strength is the stress under which the sample breaks by this bending process. It is calculated using the above stress formula, where the load p^* is the load that causes breaking. (It should be mentioned that this test is not

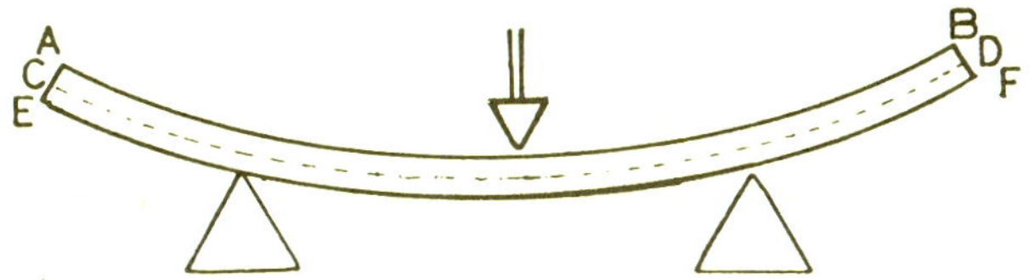

FIGURE 3.5 Sketch of stress/strain forces in sample under compression.

applicable to all plastics, since instead of breaking in the test, some of them are so flexible that the sample is forced down between the supports.)

Flexural modulus is a measure of the stiffness during the first part of the bending process. Since the word *modulus* means a "measure of," flexural modulus means "a measure of flexing or bending." In the most precise language, flexural modulus should be called "modulus of elasticity," "elastic modulus," or simply "modulus." Furthermore, for practical purposes, flexural modulus, as determined by ASTM D-790, is in many cases equivalent to tensile modulus as determined by ASTM D-638. Practical considerations indicate which method should be used.

Flexural modulus is calculated by making a graph of the data determined during the test. The graph is called a stress/strain diagram (see Figure 3.6).

The modulus is represented by the slope of the initial straight-line portion of the curve. It is calculated by dividing the change in stress *X* by the corresponding change in strain *Y*. The ASTM method refers to shortcuts that can be used in the calculations.

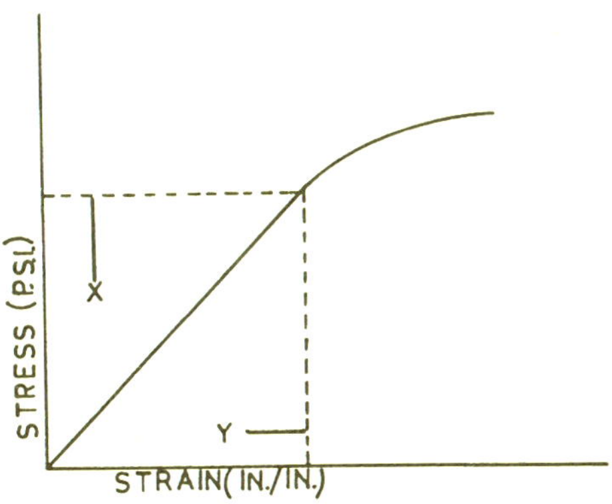

FIGURE 3.6 Sketch of typical stress/strain curve.

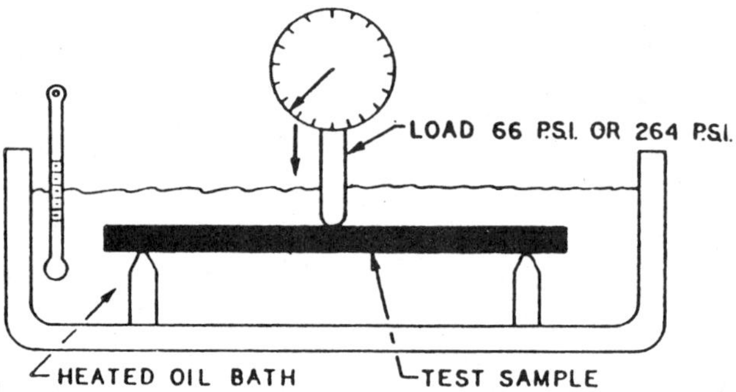

FIGURE 3.7 Sketch of heat deflection test assembly. (From Ref. 56.)

Modulus has units of pounds per square inch. Of course, the higher the values, the higher the stiffness.

In the determination of modulus, a number of errors can creep in. These are associated largely with the fact that stress/strain curves seldom have a truly straight-line portion, and considerable judgment must be used in deciding what line to draw through the data.

(a)

FIGURE 3.8 (a) Photograph of apparatus used for determining deflection temperature under load. (b) Photograph of individual testing fixture in heat deflection temperature apparatus. (From Ref. 57.)

(b)

Heat Deflection Temperature of Plastics Under Load (ASTM D-648)

The *heat deflection temperature* as determined by ASTM D-648 provides a measure of the stiffness of a plastic at elevated temperatures. At the heat deflection temperature, the polymer begins to stretch at a rapid rate over a narrow temperature interval. This temperature is near the glass transition temperature for amorphous materials, while for highly crystalline polymers it is close to the melting point. Quite often it is taken as a temperature at which the elongation reaches a particular value, such as 1%. Although it is a very specialized test, and is of questionable value, it has been used in the plastics industry for many years. In the test the sample is subjected to a load which causes it to bend, as illustrated in Figure 3.7, as temperature is increased. The sample, supports, and loading rod are all submerged in an oil bath, as described in the ASTM test procedure (see Figure 3.8). During the test, the oil is heated at a constant rate of

2°C (3.6°F)/min. The end of the test is reached when the sample has been bent 10 mils (0.01 in.) as read on an accurate gauge. The temperature of the oil at that point is called the heat deflection temperature.

The heat deflection temperature is commonly measured and reported under two conditions of loading. Actually, the load is specified in terms of stress (66 and 264 psi), rather than the actual weight of the load applied, so as to accommodate the testing of specimens with different dimensions.

The following formula relates the load, stress, and dimensions:

$$p^* = \frac{2Sbd^2}{3l} \tag{3}$$

where

p^* = load (lb)
S = stress (psi) 66 or 264
b = sample width (in.)
d = sample thickness (in.)
l = length of span between supports (in.)

The test is subject to a number of pitfalls. Perhaps the greatest is that a deflection of only 0.01 in. was established as the end point. This is an extremely small amount of movement, and factors other than the load often contribute to it. For example, warpage or stress relief of the sample during the heating process will give misleading readings; annealing of samples before testing at times may be necessary to obtain reproducible results.

The interpretation of heat deflection temperature is not at all straightforward. Superficially, it may be said that this test measures the temperature at which plastics have equivalent stiffnesses. However, this statement must be qualified to say that it is true only for these specific test conditions, and practical conditions may be different. In spite of these limitations, it continues to be used as a guide when selecting engineering plastics based on end-use temperature requirements.

The ASTM heat distortion test basically provides a single point on the deflection–temperature curve at a specific level of stress. Thus the test is particularly suitable only for quality control and selective comparisons during development work. It cannot give an indication of behavior at other stress levels and hence cannot be used for design purposes.

It has been shown, however,[48] that the single-point heat deflection temperature value obtained under ASTM D-648 conditions can be used as a normalizing factor to create heat deflection temperature-versus-stress curves for different grades of a specific type of polymer. Such curves have a distinct advantage in that further experimentation is curtailed and the heat distortion curves for the grade of interest can be generated from the master curve through the single ASTM D-648 measurement.

Figure 3.9 shows the variation of the heat deflection temperature with ap-

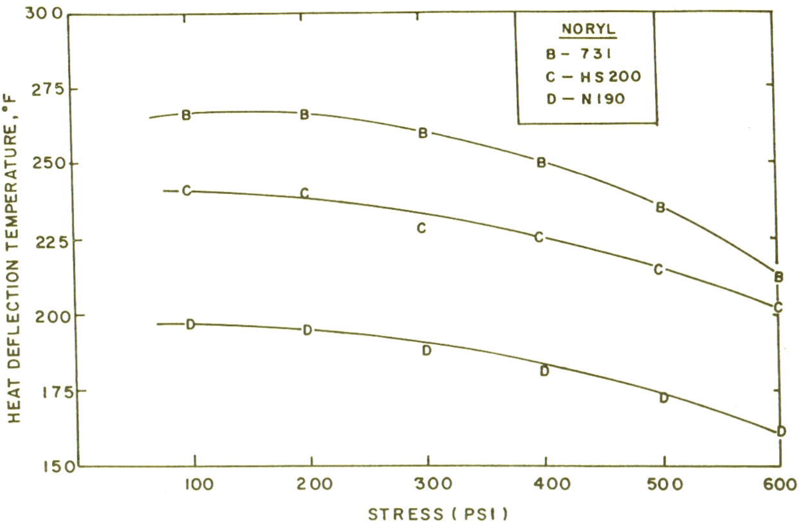

Figure 3.9 Heat distortion (deflection) temperature versus stress for Noryl®.

plied stresses for Noryl resins of three different grades. The heat deflection temperatures at 66 psi and 264 psi each represent a point on the individual curves in Figure 3.9. It is proposed that these values may be used as a normalizing factor such that revised plots are made of HDT(S)/HDT(66 psi) versus stress × HDT (66 psi) as well as HDT(S)/HDT(264 psi) versus stress × HDT(264 psi) as shown in Figures 3.10 and 3.11. It can be seen that regardless of the stress level chosen for the reference HDT used in normalization, a unified curve results from the coalescence of the data in Figure 3.9. Plots in Figures 3.10 and

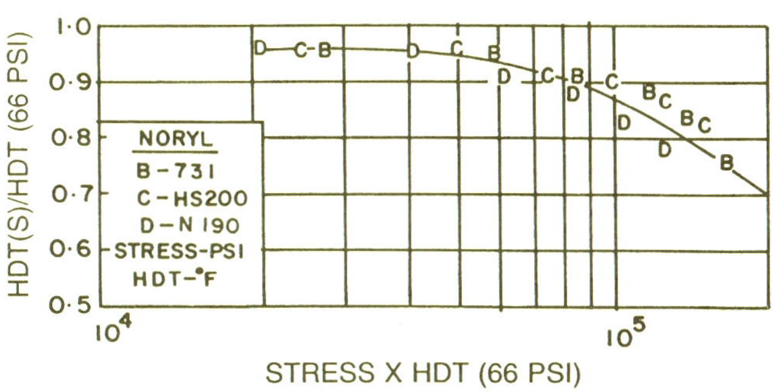

Figure 3.10 Coalesced curve for HDT versus stress that is grade invariant using HDT as the normalizing factor under a test condition of 66 psi (4.6 kg/cm²). (From Ref. 48. Reprinted with permission of Elsevier Science Ltd., Kidlington, United Kingdom.)

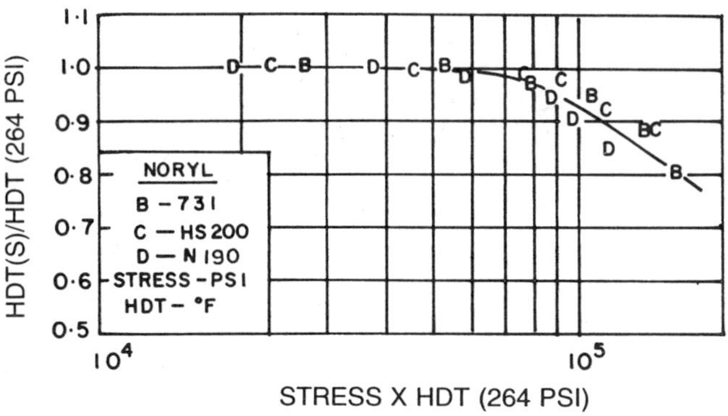

FIGURE 3.11 Coalesced curve for HDT versus stress that is grade invariant using HDT as the normalizing factor under a test condition of 264 psi (18.5 kg/cm^2). (From Ref 48. Reprinted with permission of Elsevier Science Ltd., Kidlington, United Kingdom.)

3.11 can be used for generating the heat distortion curves for any grade of Noryl merely by substituting the value of HDT determined under conditions prescribed in ASTM D 648.

Thus, once curves such as those developed in Figures 3.10 and 3.11 have been generated for various polymers, ASTM D 648 will have achieved an upgraded status, as it will then be able to provide more information than just a single-point value.

Impact Resistance of Plastics (Izod Impact Test) (ASTM D-256)

Impact resistance pertains to the ability of a material to withstand such mechanical abuse as being struck by a dropped weight or by a blow from a hammer. For practical end-use purposes, plastic parts are often evaluated by striking them with hammers, dropping them from a distance, etc.; but for evaluating or comparing, a number of plastics standard tests are used. The Izod impact test (ASTM D-256) is one of them. The test consists of breaking a test bar and calculating the amount of energy required to break it (see Figure 3.12).

The test bar measures 2½ × ½ × ⅛ in., and is notched across one edge. The notch is milled into it with a special milling cutter at a specified angle, and to a depth of 0.1 in. The specimen is clamped into the testing machine as shown in Figure 3.13.

The testing apparatus has a weighted pendulum arm that is released from a fixed height and strikes the sample in a specified manner. After the breaking

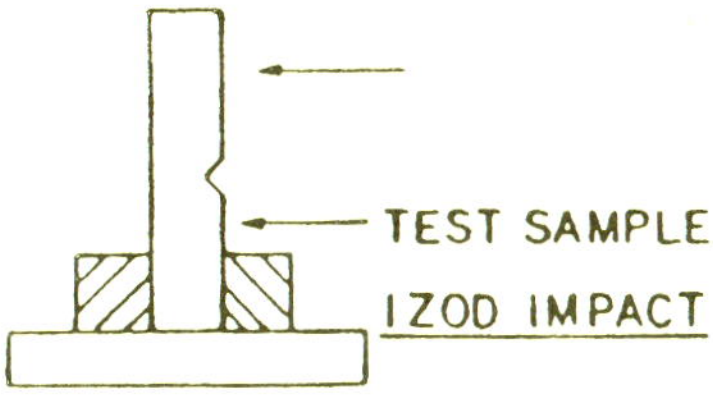

FIGURE 3.12 Sketch of Izod impact test assembly (From Ref. 56.)

of the sample, the arm follows through in the same direction and moves a pointer that gives a reading. A schematic sketch is shown in Figure 3.14.

The distance to follow-through is proportional to the amount of energy required to break the specimen. The longer the follow-through, the less energy is required to break the test piece.

The results of the test are reported in terms of foot-pounds per square inch, and are calculated from the scale readings and sample thickness measurements. Although it is common practice to use samples that are ⅛ in. thick, the method also allows for samples that are ¼ or ½ in. thick. To put all results on a standard basis, the results are reported as if all samples were 1 in. thick.

The notched Izod test has often been questioned as a meaningful measure of impact resistance because it tends to measure notch sensitivity rather than the ability of the plastic to withstand impact. Nevertheless, the test results are always included in the list of properties of a plastic, and are accepted as a guide for a toughness comparison between materials.

Creep

Creep (or deformation under load) may be defined as the deformation of a part that takes place over extended periods of time while the part is supporting a load. The sketches in Figure 3.15 illustrate this behavior.

In dealing with plastics, creep is important, since it takes place even at room temperature or below. In the design of plastic parts that are intended to support a load over a long period of time, *creep should always be considered one of the important design variables.*

There is no established method of determining creep. The ASTM test D-674 describes a method that is frequently used, although it is actually not a test method but a "recommended practice for creep tests." It discusses the complications of measuring creep, and the precautions to be taken in using creep data. The complications are due largely to the fact that the creep measurements are made over a long period of time—several months to a year or more. ASTM recognizes that creep tests are research tests by nature and not routine.

Briefly, creep tests are carried out by placing test bars in a constant-temperature chamber and attaching a load to the sample, either directly or through a level system for multiplying the load. Then, very careful measurements of the deflection or extension of the sample are made over a period of time, starting with hourly, then daily, and finally weekly measurements. Usually, specially built micrometers are used for measuring the small changes that are of interest.

(a)

FIGURE 3.13 (a) Apparatus for Izod impact test. (b) Close-up view of sample and striking arm in Izod impact test. (From Ref. 57.)

(b) 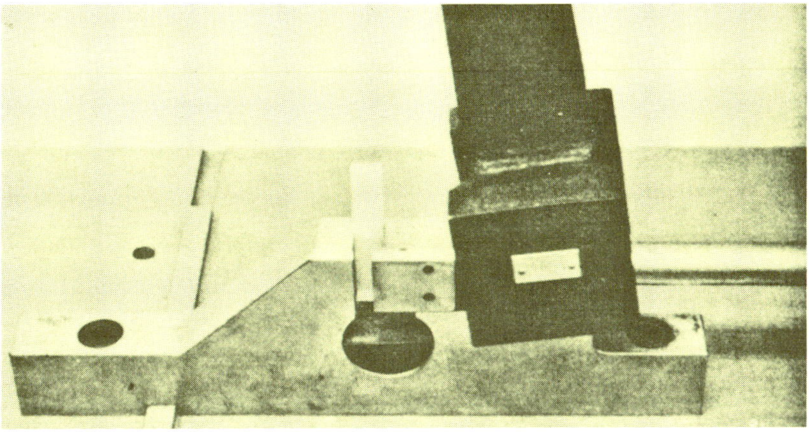

The original data from creep evaluations are often plotted on a deformation-versus-time graph, as illustrated in Figure 3.16.

From the standpoint of fundamentals, creep measurements are carried out under constant-stress (load) conditions; and the deformation (strain) continues to increase with time. It was noted earlier in this chapter with regard to tensile and flexural tests that the data concerning the initial portions of stress/strain diagrams are often summarized by calculating a modulus, which is done by dividing the stress by the strain:

$$\text{(Modulus) } E^* = \frac{\text{stress}}{\text{strain}} \tag{3.4}$$

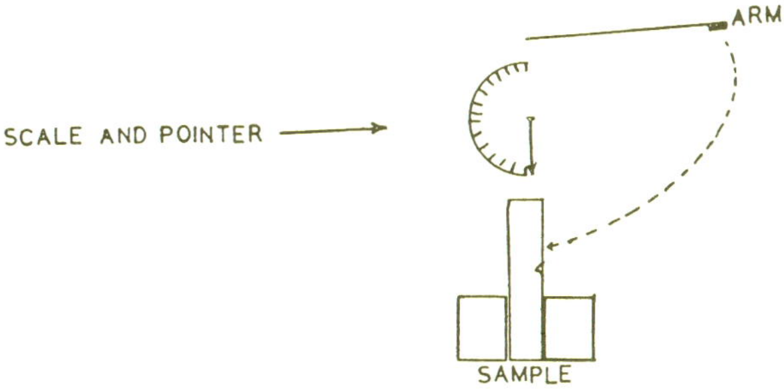

FIGURE 3.14 Schematic of Izod impact test pendulum arm.

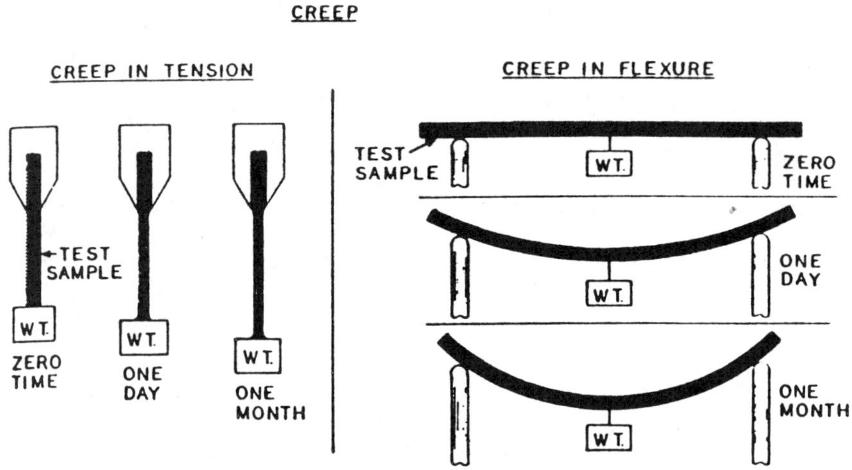

FIGURE 3.15 Schematic of creep test. (From Ref. 56.)

Similarly, creep data are put into a useful design engineering form by calculating a term called an *apparent modulus* (E_a^*). Apparent modulus is calculated in the same way, by dividing the stress by the strain; but as strain continues to increase with time, the apparent modulus decreases with time. Thus, a value of modulus applies only at a particular time.

Creep data are usually summarized in a graph of apparent modulus versus time for several levels of constant stress. An example is shown in Figure 3.17. Such data make it possible to calculate with standard engineering design formulas the deformation to be expected at a given time. It is important to ap-

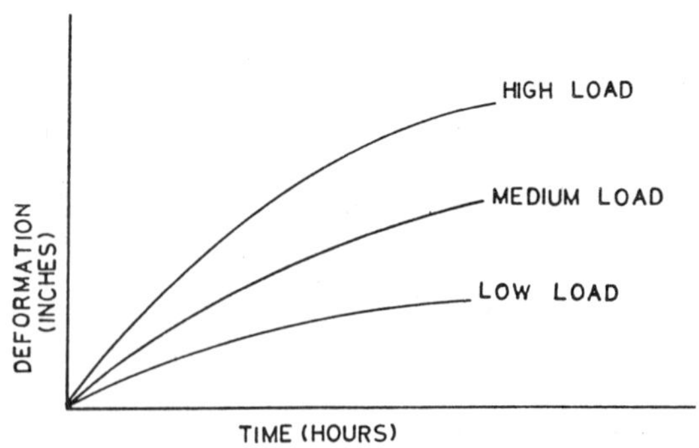

FIGURE 3.16 Example curves of deformation versus load at several load levels.

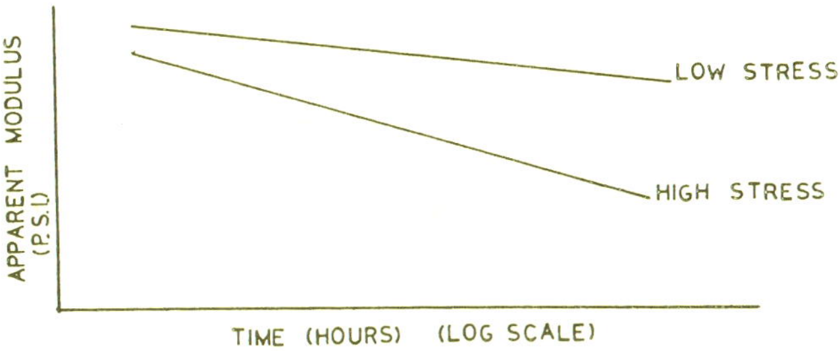

FIGURE 3.17 Diagram of apparent modulus versus time.

preciate the fact that apparent modulus is dependent on temperature, with decreasing values being obtained with increasing temperature.

Unfortunately, engineering plastics suppliers do not always use the same load or temperature conditions, or even the same ASTM test. It is therefore left to the design engineer to take careful note of which test method is used, and to utilize the information intelligently.

The cost of the experiments needed to generate stress/strain/time functions for an elapsed time of a few year is very high, and an evaluation of one sample cannot be applied to every grade or over a wide range of ambient conditions. It would thus be very attractive if a method could be evolved by which tests of shorter duration could be carried out and results carefully analyzed to predict long-term performance. An attempt was made[47] to study carefully the characteristic behavior pattern of a number of different polymers through existing stress/strain/time functions and then provide a method for coalescing the stress-versus-time curves in order to obtain master curves which could then be used for performance prediction. Though it is known that creep experiments are presented in three different ways—strain versus time (or log time), stress versus strain at various times (constant time sections across creep curves), and stress versus time or log time (constant strain section across creep curves)—only the last type of presentation was used to demonstrate the coalescence approach. Through a careful behavioral pattern study of stress-versus-log time curves of various polymers such as poly(vinyl chloride), polypropylene, acetal copolymer and polymethyl methacrylate, it was deduced that the set of curves for each polymer could be coalesced into a single master curve through a normalizing factor, namely, the stress value at an elapsed time equivalent to one day.

Figure 3.18 shows creep isometric data[58] for pipe-grade poly(vinyl chloride) (PVC) at 20°C. The data presented are for very low values of strain (0.005–0.02%), which are relevant for the final application of pipes. Each of the curves in Figure 3.18 were normalized by the respective stress value at an elapsed time of one day, σ_0, and replotted to give Figure 3.19. It can be seen that the four different curves of Figure 3.18 show an apparent regularity and fall on the single curve shown in Figure 3.19 through this simple normalizing technique.

Figure 3.20 shows the isometric stress-versus-log time curves[58] at 20°C for a polypropylene homopolymer. The data presented are for much higher strains (ranging from 0.5 to 3.0%) in comparison with the low strains in Figure 3.18. Despite this difference in the strain range, the coalescence of the curves through a plot of σ/σ_0 versus log t/σ_0 in Figure 3.21 is quite satisfactory. However, it is seen that the scatter on the master curve in Figure 3.21 is much greater than in Figure 3.19, leading to an apparent conclusion that the suggested approach is probably more effective at lower strains.

The creep data on the above two polymers were available at room temperature (20°C). It would be interesting to check the validity of the approach even at higher temperatures. For this purpose, acetal copolymer stress-versus-log time data[59] for median strain range (0.5–1.5%) at 60°C were considered (Figure 3.22). Again, the plot of σ/σ_0 versus t/σ_0 is seen to give a good coalesced curve, as can be seen from Figure 3.23.

The polymethyl methacrylate data[58] shown in Figure 3.24 were used to establish the propriety of the technique to be independent of data temperature. Figure 3.25 shows how the coalescence is rather effective, even for data from 20 to 80°C.

Figure 3.26 shows a set of curves giving comparative stress versus the stress time for the internal pressure in poly(vinylidene fluoride) (PVDF) pipes of a specific grade of DYFLOR 2000. This figure is taken from the technical brochure of Dynamit Nobel on DYFLOR 2000. It can be seen that coales-

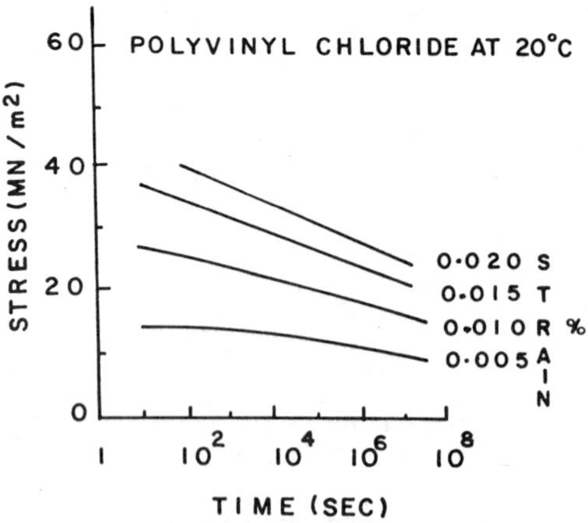

FIGURE 3.18 Isometric stress-versus-log time curves for pipe-grade PVC at 20°C under various low-strain conditions (From Ref. 58.)

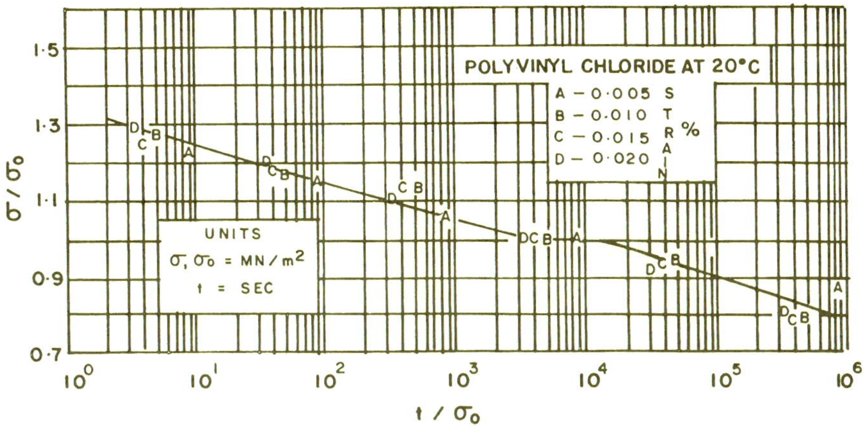

Figure 3.19 Coalesced plot of normalized stress versus normalized time for PVC at 20°C under various low-strain conditions using the stress value corresponding to one day, σ_0, as the normalizing factor. (From Ref. 47. Reprinted with permission of Elsevier Science Ltd., Kidlington, United Kingdom.)

cence of the curves[60] through a plot of σ/σ_0 versus t/σ_0 in Figure 3.27 is quite satisfactory.

In each of the coalesced curves given in Figures 3.19, 3.21, 3.23, 3.25, and 3.27, it is seen that the general trend is the same. Essentially, there exist two straight lines joined by a plateau at $\sigma/\sigma_0 = 1$. This plateau is seen to occur in the time-scale value of t/σ_0 from 10^3 to 10^5. The scatter of points below $t/\sigma_0 = 10^3$ is seen to be greater, but this region is of less relevance as it corre-

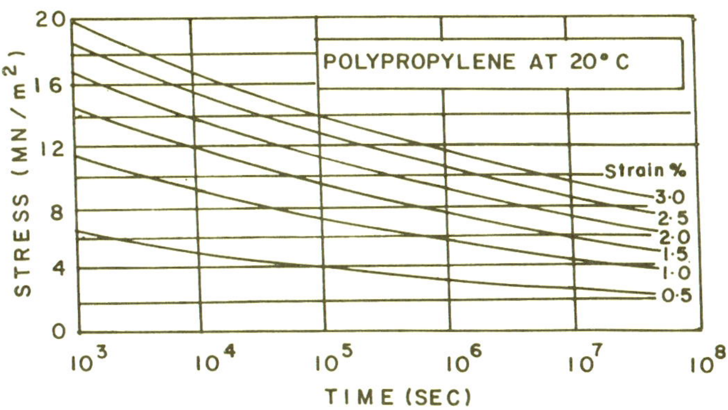

Figure 3.20 Isometric stress-versus-log time curves for polypropylene homopolymer at 20°C under various high-strain conditions. (From Ref. 58.)

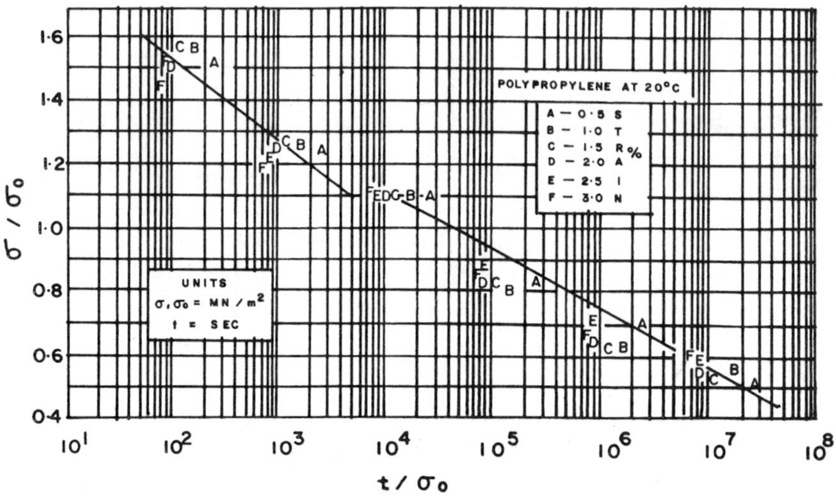

FIGURE 3.21 Coalesced plot of normalized stress versus normalized time for polypropylene homopolymer at 20°C under various high-strain conditions using the stress value corresponding to one day, σ_0, as the normalizing factor. (From Ref. 47. Reprinted with permission of Elsevier Science Ltd., Kidlington, United Kingdom.)

sponds to a time scale of less than a day. In the region of $t/\sigma_0 > 10^4$, the master curve has been drawn through the points so that the maximum scatter is below the line. This way, if a value of σ_0 is determined and used for regenerating the relevant curve of interest from the master curve, it would only result in an overdesign.

Fatigue

Fatigue relates to the failure of materials upon being subjected to repeated loads. These repeated loads may be bending, stretching, twisting, or compressing. After being subjected to such forces for many cycles (hundreds, thousands, or even millions), the material may break under a considerably lower load than that which would cause failure on a single loading. Thus, fatigue relates to the material becoming tired and weaker after cyclic distortion.

However, fatigue in plastics is not that simple. Plastics do not just become weaker and weaker under *all* conditions of loading. At high loads they do; but at low loads, some can be loaded repeatedly, "forever," without any loss of strength. It then becomes the purpose of fatigue tests to determine the highest level of repeated loading that a plastic can withstand in long-term use. Such a level of loading is called a *fatigue endurance limit*. The term has been borrowed

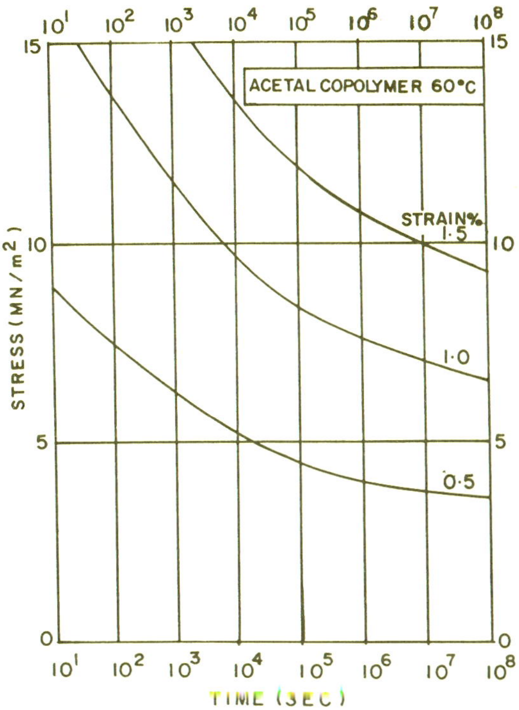

FIGURE 3.22 Isometric stress-versus-log time curves for acetal copolymer at 60°C under various medium-strain conditions. (From Ref. 59. Reprinted with permission of Chapman & Hall, Andover, United Kingdom.)

from metallurgical and mechanical engineers, since this property was first studied in metals. Some scientists question its use for plastics; but until it is replaced by a more useful term, it will probably remain in use.

Basically, fatigue tests consist of evaluating a series of samples under different loads. Samples subjected to a high load fall in a short time (few cycles), and those under lower loads after a long time (many cycles). A plot of the data is made as in the example sketch in Figure 3.28. It can be seen that the curve levels off, and that a stress can be determined with reasonable certainty that should not be exceeded for long-term repeated loading.

Considering tension and compression alone, there remain a number of general conditions. The sample may be pulled in tension only, or the load may cycle from one of tension to one of compression. The sample may be preloaded. For example, it could be preloaded with a low load in tension and cycled to a high load in tension.

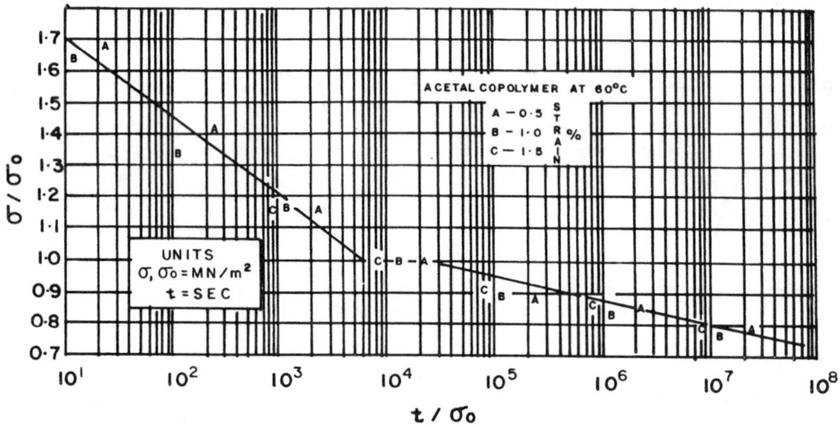

FIGURE 3.23 Coalesced plot of normalized stress versus normalized time for acetal copolymer at 60°C under various medium-strain conditions using the stress value corresponding to one day, σ_0, as the normalizing factor. (From Ref. 47. Reprinted with permission of Elsevier Science Ltd., Kidlington, United Kingdom.)

The test must also be defined with respect to whether it is a constant-load or a constant-deformation test. In the first case, with constant load, the deformation would continue to increase. In the second case, when constant deformation is the objective, the load would continuously decrease.

These complications are mentioned to point out that there is no one stan-

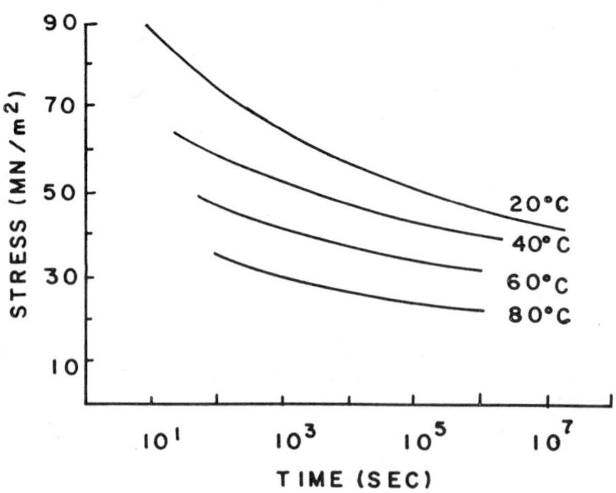

FIGURE 3.24 Isometric stress-versus-log time curves for polymethyl methacrylate at various temperatures. (From Ref. 58.)

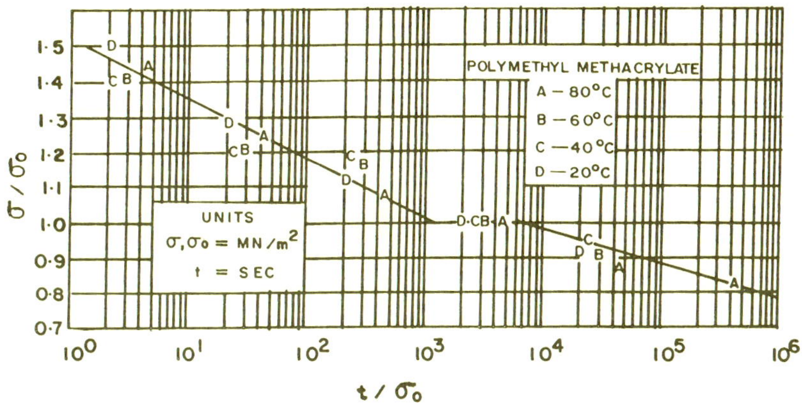

Figure 3.25 Coalesced plot of normalized stress versus normalized time for polymethyl methacrylate at various temperatures using the stress value corresponding to one day, σ_0, as the normalizing factor. (From Ref. 47. Reprinted with permission of Elsevier Science Ltd., Kidlington, United Kingdom.)

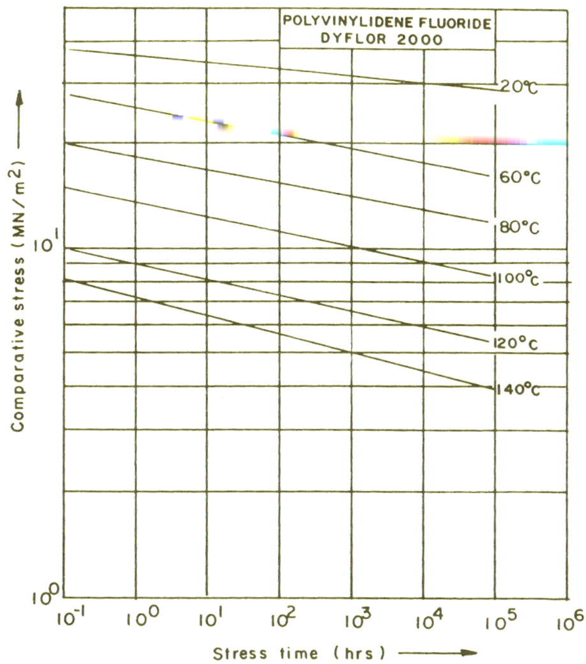

Figure 3.26 Comparative stress-versus-stress time curves at six different temperatures for internal pressure of pipe made from DYFLOR 2000. (From technical brochure of Dynamit Nobel.)

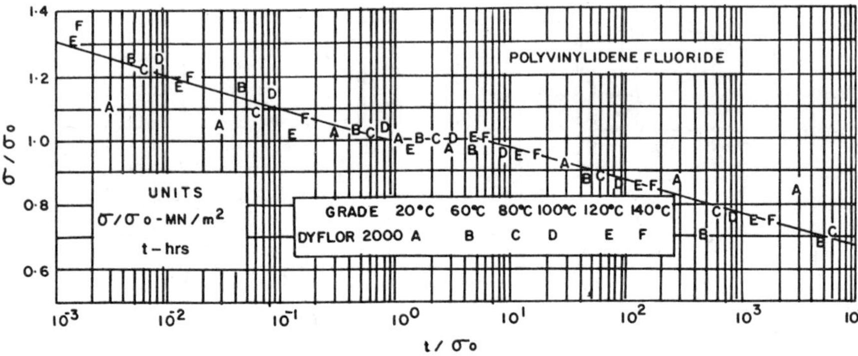

FIGURE 3.27 Coalesced plot of normalized stress versus normalized time using the stress value corresponding to 24 h as the normalizing factor. (From Ref. 60. Reprinted with permission of American Chemical Society, Washington, D.C.)

dard fatigue test covers all situations, and the conditions used in this test should be fully understood before applying fatigue data to a new application. Again, as in the case of some of the tests discussed previously, the fatigue test as reported in the literature is indeed a consideration in the resin selection process.

ASTM test D-671, frequently used to determine fatigue, is entitled "Repeated Flexural Stress (Fatigue) of Plastics." It describes a testing machine and provides a procedure for a constant-deflection test. The test procedure includes a discussion of fatigue testing, and specifically points out that it is a *research*-type test.

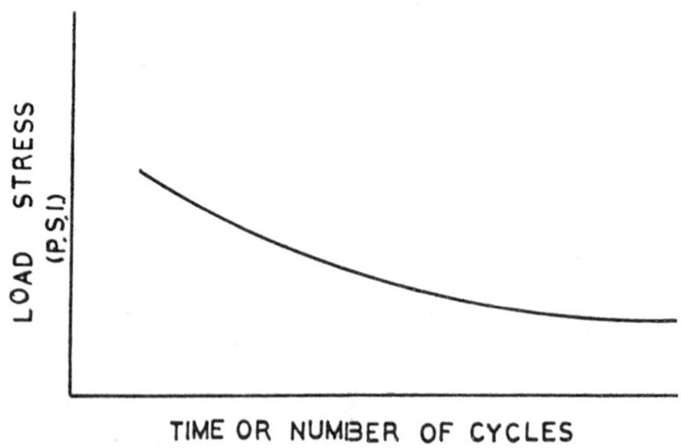

FIGURE 3.28 Diagram of fatigue test load stress versus time.

The fatigue data as reported in the literature for some resins show four cycle levels, but not all material suppliers do this. In fact, it is unfortunate that not enough data have been published to permit comparisons of all the resins discussed later in this book, since it can be a very important design consideration. It is for this reason that the test method has been covered, even though fatigue data are not compared in the property rankings in a later chapter.

Arc Resistance (ASTM D-495)

In the introduction to the test procedure itself, this ASTM standard test method calls attention to the limited conclusions that can be drawn from the test results. For example, the standard states that it will not usually permit conclusions to be drawn about the relative arc resistance rankings of materials which might be subjected to types of arcs not tested by the procedure, such as high voltage at high currents or low voltage at low or high currents. However, the test is useful for preliminary screening of materials, which is the objective of this book.

The test procedure is an involved one, requiring very specific items of test equipment, and is too complex for the intended purpose of this discourse. For details of the test, reference should be made to the ASTM D-495 procedure itself.

Basically, two electrodes are applied at specific locations on a test sample; and high voltage at low current is carefully applied until some kind of failure occurs (see Figure 3.29). Some organic compounds burst into flame without the formation of a visible conducting path in the substance. An example of one of these is an acetal. Other organic compounds fail by "tracking," that is, when a thin wiry line is formed between the electrodes. Still another type of failure occurs by carbonization of the surface until enough carbon is present to carry the current.

The results of the test are reported in seconds measured until failure.

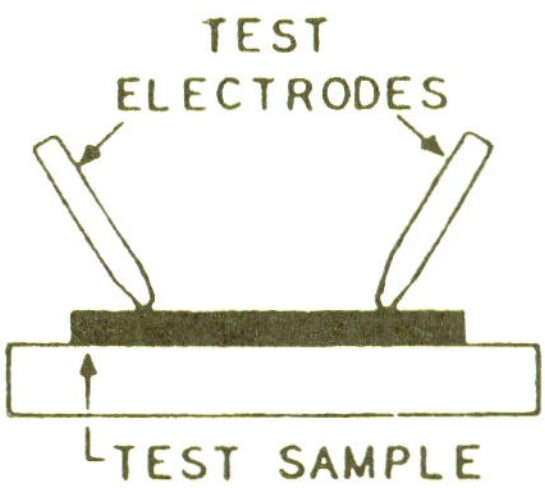

Figure 3.29 Sketch of arc resistance test assembly. (From Ref. 56.)

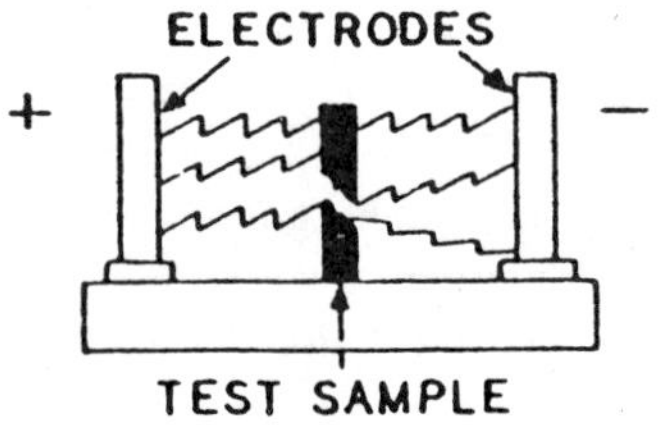

FIGURE 3.30 Sketch of dielectric strength test assembly. (From Ref. 56.)

Dielectric Strength (ASTM D-149)

Dielectric Strength is defined as the ratio of the dielectric breakdown voltage to the thickness of an insulating material. The test procedure states that when determining the dielectric breakdown voltage of a material, any one of three different methods may be employed for applying test voltage. These are the short-time test for quick determination, and the step-by-step test and the slow-rate-of-rise test, where deviation of voltage is important. In choosing the type of test, reference should always be made to the ASTM method applicable to the material being tested.

All material suppliers have not chosen the same method, making the ultimate ranking of materials by dielectric strength per se inexact; but the relative position of the dielectric strength of various compositions has some merit.

As is the case with arc resistance, the test procedure and equipment used for it are complex (see a schematic of the test setup in Figure 3.30), and a detailed description is outside the scope of this discussion. However, some basic principles should be understood.

For example, it should be understood that the dielectric strength is greatly dependent on the thickness of the test specimen. Basically, it varies inversely as a fractional power of the specimen thickness. Most resin suppliers provide the dielectric strength of their materials based on tests made on samples 125 mils thick. Results reported on films are not very useful in considering engineering plastics applicability in an injection-molded part.

The dielectric strength is reported in terms of volts per mil of sample material.

4

Rheological Property Data

INTRODUCTION

Rheology is concerned with the description of the deformation of material under the influence of stresses. Deformation and flow exist in the processing of polymer melts.[61] The various processing techniques employed in converting plastics into finished products are given in Table 4.1.[62] It is through these various processing techniques that a myriad of products can be formed for various applications.

Millions of tons of polymers are processed each year all over the world. However, not all of the material is formed into final products for commercial benefit. The reason for lower material yield is that the processor is unable to maintain product quality due to ignorance regarding the process and raw material. What is important is to get to know the material and the process well. This

TABLE 4.1 Capsule Review of Selected Plastics Processing Methods

Process	Description	Key advantages	Notable limitations
Injection molding	Thermoplastic raw material is heated to plasticity in cylinder at controlled temperature, then forced under pressure through a nozzle into sprues, runners, gates, and cavities of mold. The resin solidifies rapidly, the mold is opened, and the part ejected. In modified versions of process—runnerless molding—the runners are part of the mold cavity.	Extremely rapid production rates, hence low cost per part; little finishing required; good dimensional accuracy; ability to produce relatively large, complex shapes; very good surface finish.	High initial tool and die costs; not practical for small runs.
Compression molding	Thermoplastic raw material is placed in a heated mold cavity; the mold is closed, heat and pressure are applied, and the material flows and fills the mold cavity.	Little waste of material and low finish costs; large, bulky parts possible.	Extremely intricate parts involving undercuts, side draws, small holes, delicate inserts, etc., not practical; very close tolerances difficult to produce.
Calendering	Doughlike thermoplastic mass is worked into a sheet of uniform thickness by passing it through and over a series of heated or cooled rolls. Calendars are also used to apply plastic covering to other materials.	Low cost; sheet materials are virtually free of molded-in stress—i.e., they are isotropic.	Limited to sheet materials; very thin films not possible.
Extrusion	Thermoplastic raw material is fed from a hopper to a screw and barrel, heated to plasticity and then forwarded, usually by a rotating screw, through a nozzle having the desired cross-section configuration.	Low tool cost; a great many complex profile shapes possible; very rapid production rates; can apply coatings or jacketing to core materials, such as wire.	Limited to sections of uniform cross section.
Thermoforming	Heat-softened thermoplastic sheet is placed over male or female mold. Air is evacuated from between the sheets and mold, causing the sheet to conform to the contour of the mold. There are many variations, including vacuum snapback, plug assist, drape forming, etc.	Tooling costs are generally low; produces large parts with thin sections; often economical for limited production of parts.	Limited to parts of simple configuration and to a narrow choice of materials; produces high quantities of scrap.
Blow molding	An extruded tube (parison) of heated thermoplastic is placed between two halves of an open-split mold and expanded against the sides of the closed mold by air pressure. The mold is opened, and the part ejected.	Low tool and die cost; rapid production rates; ability to mold relatively complex hollow shapes in one piece.	Limited to hollow or tubular parts; wall thickness and tolerances often hard to control.
Rotational molding	A predetermined amount of powdered or liquid thermoplastic raw material is poured into the mold. The mold is closed, heated, and rotated in the axis of two planes until the contents have fused to the inner walls of mold. The mold is opened and the part removed.	Low mold cost; large, hollow parts in one piece can be produced; molded parts are essentially isotropic in nature.	Limited to hollow parts; in general, production rates are slow.

Source: Ref. 62.

is why it is a must to understand the chemistry, physics, and engineering of polymers. The chemistry of polymers gives insight into the material properties of the polymers, which is absolutely essential for selection of the right polymer type and grade for a particular product with specific property targets. The physics and engineering of the polymers are essential when, having decided upon the right polymer and grade, the polymer chips or powder are to be converted into final product. The general concept is that polymer scientists and engineers should know the chemistry, physics, and engineering of the polymers, while the polymer processor should know how to make useful and durable products at minimum cost. For maximizing profit it is important to improve product quality, minimize material waste and production downtime, and optimize the process. All these are within the control of the polymer processor.

Polymer processing is not an art as looked upon by the entrepreneur; rather, it is the science of the flow and deformation of polymeric materials, namely, rheology. Polymer processing operations such as injection molding, compression molding, calendering, and extrusion involve deformation of polymeric material over a wide range of temperatures (170–320° C) and a broad range of shear rates (see, for example, Table 4.2). In the extruder barrel and in the runner of an injection unit, shear rates are of the order of 10 to 100 s^{-1}. The temperatures involved in the flow deformation also vary with the process. Thermoforming, for example, requires relatively low processing temperatures as compared to extrusion and injection molding. Thus, what the processor requires is knowledge of the flow behavior of the melt at various temperatures relevant to processing and over a wide range of shear rates of about four decades. Flow information basically consists of data on the viscosity and elasticity of the melt.

COMPLEXITIES IN RHEOLOGICAL DATA GENERATION

Within the family of resins of a particular chemical type, a large number of grades are available to the product designer. Thus, flow data have to be generated for each grade the processor intends to use within a generic type of poly-

TABLE 4.2 Shear Rate Ranges Encountered in Common Polymer Processing Operations:

Process	Typical shear rate (s^{-1})
Compression molding	1–10
Calendering	$10–10^2$
Extrusion	$10^2–10^3$
Injection molding	$10^3–10^4$

Source: Ref. 54. (Reprinted with permission of Gulf Publishing Company, Houston, Texas.)

mer. Figure 4.1 shows the variation in the viscosity of polyether sulfone (PES) Victrex® 200P melt at different temperatures over a wide range of shear rates.[25] This flow information is useful only if the particular grade of Victrex® 200P PES is being processed. However, a number of grades of PES are available, for example, Victrex® 300P, which responds differently to deformation. Thus, if the processor uses different grades during a manufacturing process, then curves of the type shown in Figure 4.1 are essential for designing the processing equipment, process optimization, and troubleshooting. Figure 4.2 gives another example of necessary rheological data.[25] Generation of flow data of the type shown in Figures 4.1 and 4.2 requires highly sophisticated and expensive rheological equipment.

It is not only the financial burden of buying and maintaining the equipment that is difficult, but the fact that it is time consuming, cumbersome, and requires trained operators, leaving no incentive for polymer processors to use rheology as a tool for increasing productivity and product quality. Further, interpretation of the generated data requires good technical backing and the need to employ extra technical staff for the polymer processor, which is not desirable from the entrepreneur's point of view. Thus the cost of making fundamental investigations into polymer properties and producing basic rheological data of the variation of viscosity with shear rate and temperature on the large number of available grades is very high.

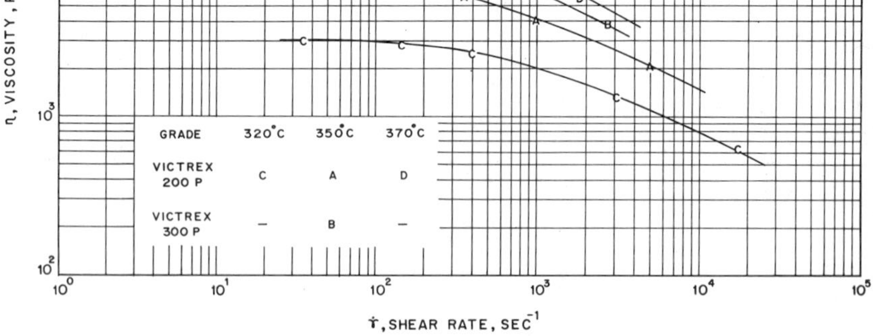

FIGURE 4.1 Viscosity-versus-shear rate plots at different temperatures for two different grades of PES with different MFIs.

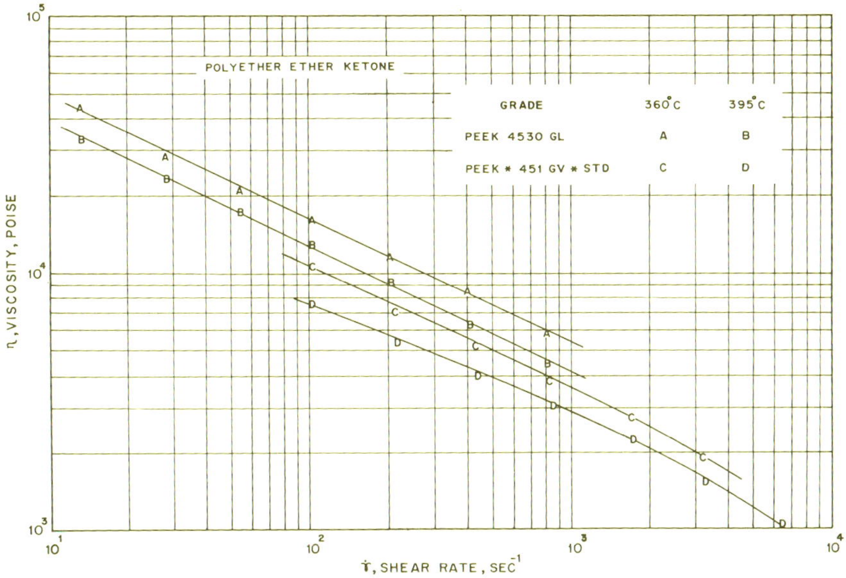

FIGURE 4.2 Viscosity-versus-shear rate plots at two different temperatures for two different grades of PEEK with different MFIs.

A SIMPLIFIED APPROACH TO RHEOLOGY

Many polymer processors are aware that rheology can serve as a handy aid to provide pointers to processing problems. As a compromise, they attempt to obtain flow information on a comparative basis through a simple melt flow apparatus. This apparatus is easy to operate and cheap and provides single-point values of the viscosity of the polymer sample in terms of melt flow index. *Melt flow index* (MFI) is defined as the weight of the polymer extruded in grams in 10 min from a melt flow apparatus under conditions of standard load and temperature as specified in ASTM D1238.[63] The geometric parameters of the melt flow indexer and a description of the test conditions for commonly used polymers are available in the standard.[63]

Although MFI is a good indicator of the most suitable end use for which a particular grade can be used, it is not a fundamental polymer property. It is an empirically defined parameter that is critically influenced by the conditions of measurement, besides the physical properties and molecular structure of the polymer. Since the values of temperatures and shear rate employed in this test differ substantially from those encountered in actual large-scale processes, MFI test results do not correlate directly with processing behavior.

Literature does not encourage the use of MFI for obtaining guidelines in processing and insists that an entire flow curve or rheogram is essential if one desires to use rheology as an aid to polymer processing.[64,65] For most design and optimization purposes, an order-of-magnitude estimate of the viscosity is adequate.

A method was successfully developed for generating the complete flow curve at different temperatures for a resin from just the knowledge of its MFI.[66] The strategy was to investigate the dependence of the rheograms of various grades (as specified by MFIs) of the same polymer on temperature and dead weight of the MFI test, and then to eliminate these dependencies to produce a unified flow curve in terms of MFI.

The mechanistic rationale for unifying the curves is as follows: Since the melt flow apparatus is basically an extrusion rheometer and the MFI is basically a flow rate, conventional expression for shear stress τ and shear rate $\dot{\gamma}$ can be used to obtain the relationship between MFI and viscosity as well as shear rate. Thus,

$$\tau = \frac{R_N\, F}{2\pi R_P^2\, 1_N} \tag{4.1}$$

$$\dot{\gamma} = \frac{4Q}{\pi R_N^3} \tag{4.2}$$

where piston radius R_p = 0.4737 cm, nozzle radius R_N = 0.105 cm, nozzle length l_N = 0.8 cm, force F = test load L(kg) × 9.807 × 10 ^{5} dynes, and flow rate Q = MFI/600 cm^3/s.

Equations (4.1) and (4.2) can be simplified through proper substitutions of the fixed geometric parameters of the melt flow apparatus to give

$$\tau = 9.13 \times 10^4 \text{ L} \tag{4.3}$$

$$\dot{\gamma} = \frac{1.83 \text{ MFI}}{\rho} \tag{4.4}$$

As the MFI value is generated at a fixed temperature and a fixed load, a single point on the shear stress-versus-shear rate curve at that specific temperature can be obtained from Eqs. (4.3) and (4.4). One could make use of this fact for calculating the value of MFI from a known shear stress-versus-shear rate curve when the MFI value is not reported.

Now, using the relationship

$$\eta = \frac{\tau}{\dot{\gamma}} \tag{4.5}$$

the following expressions can be easily developed:

$$\eta \bullet \text{MFI} = 4.98 \times 10^4 \rho \text{L} \tag{4.6}$$

$$\frac{\dot{\gamma}}{\text{MFI}} = \frac{1.83}{\rho} \tag{4.7}$$

For a given polymer the density and testing load condition are fixed, thus indicating that the MFI of the material is directly proportional to the apparent shear rate and inversely proportional to the apparent viscosity under the prescribed test conditions of temperature. Although Eqs. (4.6) and (4.7) are valid only at the particular MFI test condition, in effect the validity of these equations over the entire flow curve can be constituted by a change of dead-weight condition and hence the proportionality constant. It should therefore be possible to coalesce η-versus-$\dot{\gamma}$ curves of different grades of a polymer with different MFIs by plotting (η • MFI) versus ($\dot{\gamma}$ /MFI) on a log-log scale, independent of temperature if the correct MFI value corresponding to the temperature of measurement of η-versus-$\dot{\gamma}$ curve is used.

The curve in Figure 4.1 at three different temperatures for the same grade of PES (Victrex® 200P) were coalesced using appropriate MFI values to give a unified curve as shown in Figure 4.3. It can be seen that the curve for the sec-

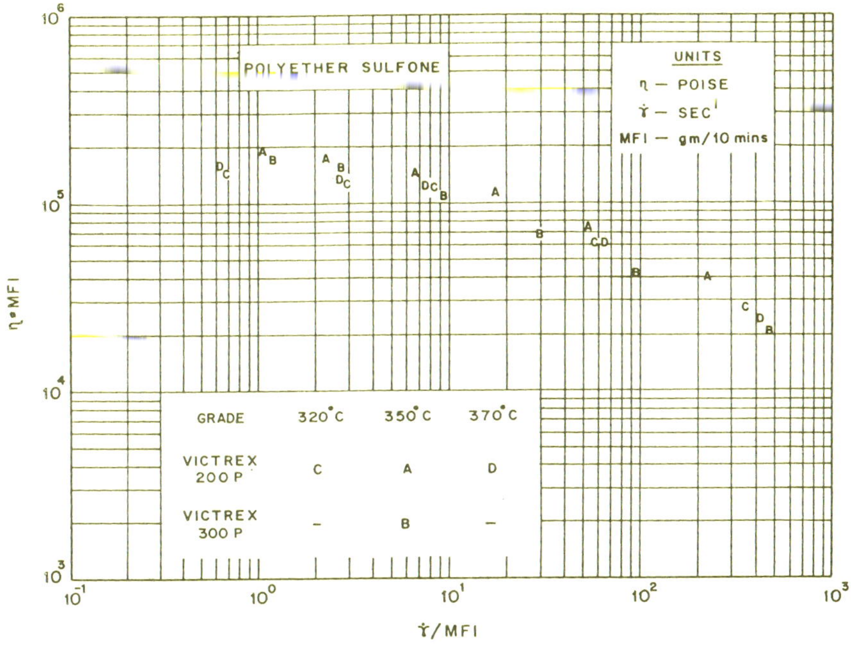

FIGURE 4.3 Master curve for PES at 5.0-kg test load condition for MFI. (From Ref. 25.)

ond grade of PES (Victrex® 300P) also coalesces on the same curve. Thus the plot shown in Figure 4.3 is independent of polymer grade and melt temperature. The curves in Figure 4.2 likewise can be coalesced to give a single master curve for PEEK, as shown in Figure 4.4. Figures 4.3 and 4.4 were obtained for a fixed MFI load condition of 5 kg. Hence, if the load condition of MFI measurement is different, the obtained master curve will be different.

Unified curves have been provided for a large number of homopolymer types, such as, engineering thermoplastics,[1] specialty polymers,[25] polyolefins and styrenics,[66,67] cellulosics,[68] vinyls,[69] and vinylidene fluorides. The technique has been shown to be effectively applicable to copolymers,[70] liquid crystalline polymers,[71] polymer blends,[72] and filled polymers.[73,74] In the case of filled polymer systems,[73] it was shown that the unified curve for systems with and without fillers is the same. In fact, even in the presence of reactive additives,[74] the unification technique gives a single curve for a particular generic type of polymer. This is illustrated by Figure 4.5, which shows the unified plot for nylons. It can be seen that data for various types of nylons along with those containing reactive additives such as lithium chloride (LiCl) all fall on one curve.

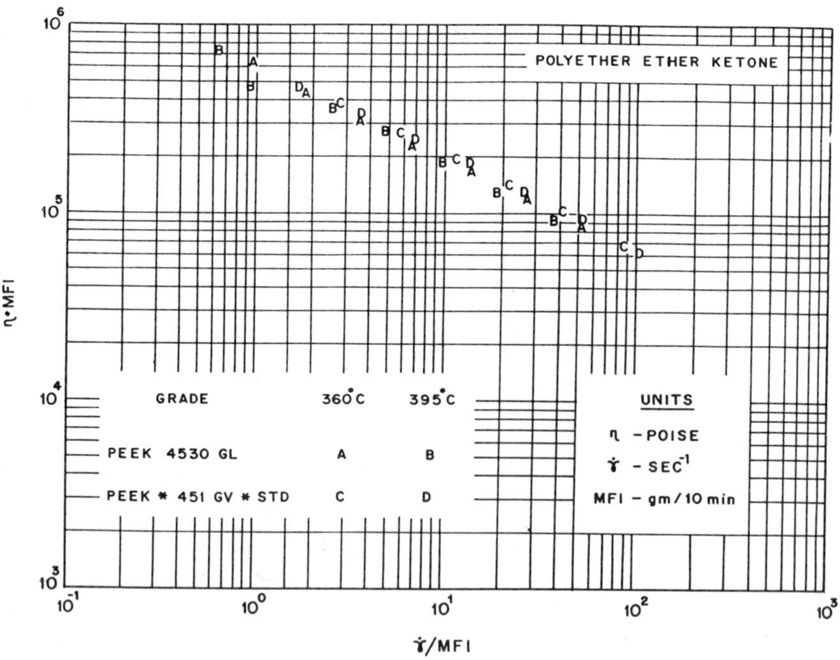

FIGURE 4.4 Master curve for PEEK at 5.0-kg test load condition for MFI. (From Ref. 25.)

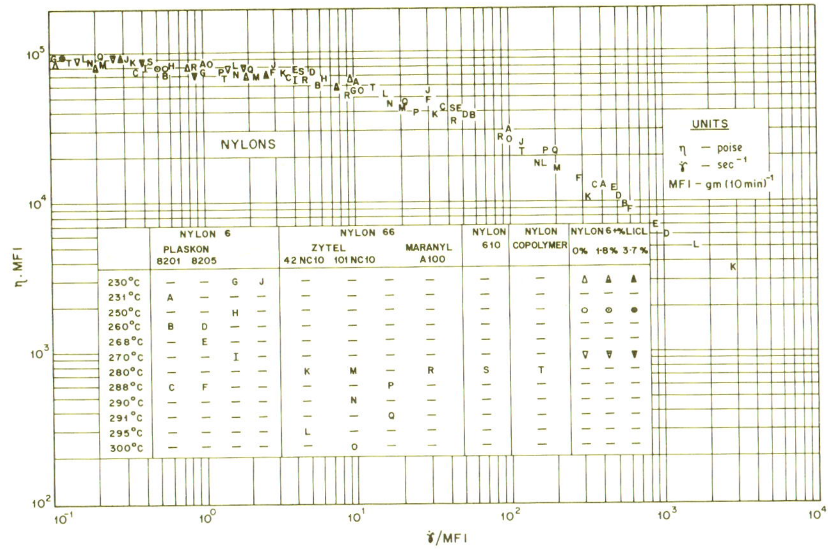

Figure 4.5 Master curve for nylons at 2.6-kg test load condition for MFI. (From Ref. 74. Reprinted with kind permission of Chapman & Hall, Andover, United Kingdom.)

A number of data points from published literature were carefully analyzed to confirm the propriety of the unified curves.[61] Generated data as well as data from different sources were used in order to account for variation in measurement techniques and the varied element of human error during measurement.[61] Considering the variables, the coalesced plots are reasonably unique and can be used for generating rheological data without the use of sophisticated, expensive equipment.

The behavior of viscosity-versus shear rate for any polymer grade at any temperature of interest can be obtained from the master curves using the following procedure.

1. First, it is essential to obtain the MFI of the polymer for the particular grade under consideration. This can be done either by making an actual measurement under standard test load and temperature conditions, or by obtaining the MFI value directly from the polymer manufacturer.
2. It may happen that the test load condition of the obtained MFI is different from that specified in the unified curve. Under such circumstances, a new MFI value should be calculated using the following equation:

$$\frac{\mathrm{MFI}_{L_2}}{\mathrm{MFI}_{L_1}} = \left(\frac{L_2}{L_1}\right)^{1/n} \tag{4.8}$$

where n is the slope of the master curve in the straight portion covering the MFI range, while L_1 and L_2 are two test load conditions.

It may also happen that the temperature at which the η-versus $\dot{\gamma}$ curve is desired is different from that at which the MFI value was determined. In such cases, one of the following two equations should be used for estimating the correct MFI value.

Modified WLF-type equation:[66]

$$\log\frac{\mathrm{MFI}_{L_2}}{\mathrm{MFI}_{L_1}} = \frac{8.86\,(T_2 - T_s)}{101.6 + (T_2 - T_s)} - \frac{8.86\,(T_1 - T_s)}{101.6 + (T_1 - T_s)} \tag{4.9}$$

Modified Arrhenius-type equation:[75]

$$\frac{\mathrm{MFI}_{T_2}}{\mathrm{MFI}_{L_1}} = \exp\left[\frac{E}{R}\left(\frac{1}{T_2} - \frac{1}{T_1}\right)\right] \tag{4.10}$$

where

T_1 = ASTM recommended test temperature, K
T_2 = temperature at which the MFI value is to be determined, K
T_s = a standard reference temperature taken to be equal to T_g + 50, K
T_g = glass transition temperature, K
R = gas constant equal to 1.9858
E = activation energy for viscous flow, values of which for a selected list of engineering thermoplastics are given in Table 4.3

The choice of which equation is to be used for determining the temperature dependence of MFI is governed mainly by whether T is less than or greater than T_g + 100 (K). At temperatures relatively nearer to T_g, it could be expected that free volume and its changes with temperature play a predominant role and hence, the WLF-type equation, Eq. (4.9), could provide better estimates. At temperatures greater than T_g + 100 (K), the temperature dependency of MFI is decisively affected by overcoming of the forces of intermolecular interactions, in which case the Arrhenius-type equation, Eq. (4.10), would give better estimates.

Once the correct MFI value is determined at the temperature of interest and under the appropriate load condition as required, then the plot of η versus γ can be readily obtained by substitution of this MFI value.

TABLE 4.3 Flow Activation Energies at a Constant Shear Stress Based on Melt Flow Index for a Selected List of Different Generic Types of Engineering Thermoplastics

Thermoplastic	E (K/cal/mol)	Temperature[a] (°C)	Test load[b] (kg)
Acrylic	38.5	170–260	3.80[c]
Nylon	18.6	230–300	2.16[c]
PET/PBT	20.1	265–285	2.16[c]
PC	21.9	250–290	1.20[c]
PVDF	8.0	190–290	12.50
PPO/PPE-based resin	33.0	260–320	5.00
PPS	4.6	280–316	5.00
PAS/PES	36.0	320–370	5.00
PEEK	12.0	360–395	5.00
PEI	30.0	355–395	5.00
PAr	35.0	288–329	5.00

[a]Applicability range of validity.
[b]Condition for MFI test.
[c]ASTM specified.
Source: Ref. 54. (Reprinted with permission of Gulf Publishing Company, Houston, Texas.)

RHEOLOGICAL MODELS FOR UNIFIED VISCOSITY-MFI CURVES

The final step in the above procedure for generating the viscosity-versus-shear rate curve from MFI involves the reading out of appropriate values from the unified curves. This could induce errors and cause a certain degree of inconvenience. In order to avoid this, appropriate rheological models have been fitted to each of the unified curves.[76,77] The forms of the rheological models are given below.

Modified Carreau model:

$$\eta \bullet \mathrm{MFI} = \eta_0 \bullet \mathrm{MFI}\left[1 + (\lambda \bullet \mathrm{MFI})^2 \bullet \left(\frac{\dot{\gamma}}{\mathrm{MFI}}\right)^2\right]^{-N} \tag{4.11}$$

Modified Ellis model:

$$\eta \bullet \mathrm{MFI} = \frac{\eta_0 \bullet \mathrm{MFI}}{1 + \left|\tau/\tau_{1/2}\right|^{a'-1}} \tag{4.12}$$

Modified Ostwald-de Waele power-law model:

$$\eta \bullet \mathrm{MFI} = K\left(\frac{\dot{\gamma}}{\mathrm{MFI}}\right)^{n-1} \tag{4.13}$$

General rheological model:

$$\eta \bullet \text{MFI} = \eta_0 \bullet \text{MFI}\left[1+\left(\frac{K}{\eta_0 \bullet \text{MFI}}\right)^P\left(\frac{\dot{\gamma}}{\text{MFI}}\right)^{(n-1)P}\right]^{1/P} \quad (4.14)$$

where

$\eta_o \bullet$ MFI = modified zero-shear viscosity function
$\eta \bullet$ MFI = modified non-Newtonian viscosity function
$\dot{\gamma}$/MFI = modified shear rate function
$\lambda \bullet$ MFI = modified time constant
K = consistency index
N, α', p and n = dimensionless parameters associated with power-law behavior
τ = shear stress given by the product of $\eta \bullet$ MFI and $\dot{\gamma}$/MFI
$\tau_{1/2}$ = value of shear stress at $\eta \bullet \text{MFI} = \eta_o \bullet \text{MFI}/2$

TABLE 4.4 Rheological Parameters of the Modified Carreau Model [Eq. (4.11)] for a Selected List of Different Generic Types of Engineering Thermoplastics

Thermoplastic	$\eta_o \bullet$ MFI [(poise) × (g per 10 min)]	$\lambda \bullet$ MFI [(s) × (g per 10 min)]	N	Modified shear rate[a] [(s^{-1}) × (g per 10 min)$^{-1}$]	Test[b] load (kg)
ABS	1.7×10^6	12.8	0.31	0.01–1	5.00[c]
Acrylic	2.1×10^5	2.4	0.16	0.1–20	3.80[c]
Nylon	9.5×10^4	11.6	0.03	0.1–20	2.16[c]
PET/PBT	8.3×10^4	0.3	0.1	0.01–50	2.16[c]
PC	4.2×10^4	0.19	0.07	1–80	1.20[c]
PVDF	6.0×10^5	0.77	0.3	0.1–4	12.50
PPO/PPE-based resin	2.2×10^5	0.2	0.32	1–40	5.00
PPS	3.4×10^5	0.46	0.36	0.001–10	5.00
PAS/PES	1.9×10^5	0.32	0.2	1–10	5.00
PEEK	4.5×10^5	1.7	0.24	0.01–20	5.00
PEI	1.4×10^5	0.038	0.33	1–200	5.00
PAr	1.8×10^5	0.04	0.38	0.01–10	5.00
Filled PP	1.2×10^5	4.6	0.13	0.1–4	2.16[c]

[a]Applicability range of $\dot{\gamma}$/MFI.
[b]Condition for MFI used in the master rheogram.
[c]ASTM specified.
Source: Ref. 61.

Tables 4.4–4.7 give the model constants for selected engineering thermoplastics based on the modified Carreau model, the modified Ellis model, the modified Ostwald-de Waele model, and the general rheological model, respectively.

The modified Carreau model and the modified Ellis model are limited to relatively low values of shear rates and shear stresses, respectively, whereas the modified Ostwald-de Waele power law model is applicable to the higher shear-rate region where the data points fall in a straight line on the log-log plot of η • MFI versus γ/MFI.

The general rheological model, however, is applicable to the entire shear rate range, covering the region of validity of the modified Carreau model as well as the region of validity of the modified Ostwald-de Waele power-law model. Hence, the general rheological model is recommended for use when the entire master rheogram is to be fitted by a single best-fitting curve.

In most industrial polymer processing operations, the shear rate ranges (as

TABLE 4.5 Rheological Parameters of the Modified Ellis Model [Eq. (4.12)] for a selected List of Different Generic Types of Engineering Thermoplastics

Thermoplastic	η_o • MFI [(poise) × (g per 10 min)]	$\tau_{1/2}$ (dyn/cm^2)	α'	Shear stress[a] (dyn/cm^2)	Test load[b] (kg)
ABS	1.7×10^6	1×10^5	2.25	10^4–10^6	5.00[c]
Acrylic	2.1×10^5	4.09×10^5	2.23	2×10^4–2×10^6	3.80[c]
Nylon	9.5×10^4	1.18×10^6	2.24	10^4–10^7	2.16[c]
PET	8.3×10^4	3×10^6	2.28	10^3–10^7	2.16[c]
PC	4.2×10^4	2.73×10^6	1.82	5×10^4–10^7	1.20[c]
PVDF	6.0×10^5	1.4×10^6	3.00	2×10^5–5×10^6	12.50
PPO/PPE-based resin	2.2×10^5	9×10^5	2.70	2×10^5–10^7	5.00
PPS	3.4×10^5	7.6×10^5	2.70	10^3–2×10^6	5.00
PAS/PES	1.9×10^5	2.2×10^6	2.10	10^5–10^7	5.00
PEEK	4.5×10^5	2.8×10^5	2.10	5×10^4–10^7	5.00
PEI	1.4×10^5	3.6×10^6	2.56	2×10^5–2×10^7	5.00
PAr	1.8×10^5	2.3×10^7	1.50	10^4–10^7	5.00
Filled PP	1.2×10^5	1.54×10^5	2.89	6×10^3–10^6	2.16[c]

[a]Applicability range for τ.
[b]Condition for MFI used in the master rheogram.
[c]ASTM specified.
Source: Ref. 61.

TABLE 4.6 Rheological Parameters of the Modified Ostwald-de Waale Power Law Model [Eq. 4.13)] for a Selected List of Different Generic Types of Engineering Thermoplastics

Thermoplastic	K [(g/cms^{2-n}) × (g per 10 min)^n]	n	Modified shear-rate[a] [(s^{-1} × (g per 10 min)$^{-1}$)]	Test load[b] (kg)
ABS	3.5 × 10^5	0.38	1–100	5.00[c]
Acrylic	3.2 × 10^5	0.44	20–1000	3.80[c]
Nylon	3.4 × 10^5	0.44	50–1000	2.16[c]
PET/PBT	5.0 × 10^5	0.43	50–1000	2.16[c]
PC	2.5 × 10^5	0.52	80–1000	1.20[c]
PVDF	7.5 × 10^5	0.40	2–200	12.50
PPO/PPE-based resin	1.0 × 10^6	0.25	40–1000	5.00
PPS	6.0 × 10^5	0.44	10–100	5.00
PAS/PES	3.0 × 10^5	0.60	10–1000	5.00
PEEK	6.0 × 10^5	0.32	20–1000	5.00
PEI	1.4 × 10^6	0.34	200–1000	5.00
PAr	2.4 × 10^5	0.85	10–100	5.00
Filled PP	1.75 × 10^5	0.34	4–1000	2.16[c]

[a]Applicability range for $\dot{\gamma}$/MFI.
[b]Condition for MFI used in the master rheogram.
[c]ASTM specified.
Source: Ref. 61.

given in Table 4.2) fall within the domain of the modified Ostwald-de Waale power law model. Hence, this model has been used more frequently, especially for the determination of various important processing parameters, from merely the knowledge of the MFI of the material.[78–83] This is done through the use of its shear stress-based form, namely,

$$\tau = (\eta \bullet \text{MFI}) \bullet \frac{\dot{\gamma}}{\text{MFI}} \tag{4.15}$$

which for Eq. (4.13) takes the following form:

$$\tau = K\left(\frac{\dot{\gamma}}{\text{MFI}}\right)^n \tag{4.16}$$

PRACTICAL APPLICATIONS OF THE UNIFICATION TECHNIQUE

The unified curves, when fitted with appropriate rheological models, provide an easy way to obtain viscosity-versus-shear rate data through mathematical calcu-

TABLE 4.7 Rheological Parameters of the General Rheological Model [Eq. (4.14)] for a Selected List of Different Generic Types of Engineering Thermoplastics

Thermoplastic	$\eta_0 \cdot$ MFI [(poise) × (g per 10 min)]	K [(g/cm·s^{2-n}) × (g per 10 min)^n]	n	P	Modified shear rate[a] [(s^{-1}) × (g per 10 min)$^{-1}$]	Test load[b] (kg)
ABS	1.7×10^6	3.5×10^5	0.38	−1.090	0.01–100	5.00[c]
Acrylic	2.1×10^5	3.2×10^5	0.44	−1.336	0.1–1000	3.80[c]
Nylon	9.5×10^4	3.4×10^5	0.44	−1.624	0.1–1000	2.16[c]
PET/PBT	8.3×10^4	5.0×10^5	0.43	−2.031	0.01–1000	2.16[c]
PC	4.2×10^4	2.5×10^5	0.52	−2.282	1–1000	1.20[c]
PVDF	6.0×10^5	7.5×10^5	0.40	−3.106	0.1–200	12.50
PPO/PPE-based resin	2.2×10^5	1.0×10^6	0.25	−1.318	1–1000	5.00
PPS	3.4×10^5	6.0×10^5	0.44	−1.306	0.001–100	5.00
PAS/PES	1.9×10^5	3.0×10^5	0.60	−4.033	1–1000	5.00
PEEK	4.5×10^5	6.0×10^5	0.32	−1.264	0.01–1000	5.00
PEI	1.4×10^5	1.4×10^6	0.34	−2.060	1–1000	5.00
PAr	1.8×10^5	2.4×10^5	0.85	−12.127	0.001–100	5.00
Filled PP	1.2×10^5	1.75×10^5	0.34	−1.076	0.1–1000	2.16[c]

[a]Applicability range for γ/MFI.
[b]Condition for MFI used in the master rheogram.
[c]ASTM specified.
Source: Ref. 61.

lations, knowing the value of MFI under correct temperature and load conditions. The other advantage of fitting the rheological models to the unified curves lies in the fact that calculations involving pressure distributions, mold design, die design, heat transfer and so on are greatly simplified,[78–83] for example,

The minimum pressure drop during cavity filling as well as the minimum clamping force to prevent mold opening in an injection-molding operation[78]
The compacting force in a compression-molding operation[79]
The pressure distribution, torque, and power input to each roller in a calendering operation[80]
The pressure losses through dies of complex cross section during extrusion[81]
The viscous heat dissipation during various processing operations[82]
Selection of the component polymers, their appropriate grades, and their composition based on property requirements, as well as the proper choice of compounding method and conditions during polyblending[83]

can all be determined rather easily using the unification technique. Details of how this is done are available in Ref. 61.

Only the aspects relating to polyblending are treated further in this chapter as they are relevant to the subject matter of this book.

Polyblending, modification of one polymer with another, involves the critical steps of selection of the component polymers, their appropriate grades, and their composition based on property requirements, as well as the proper choice of compounding method and conditions. The component grade selection and choice of compounding process parameters are often done arbitrarily based on evaluation of end properties. This involves a sort of trial-and-error procedure before finalizing the appropriate grades of the components and the compounding conditions. Synergistic property advantages cannot be achieved even in thermodynamically miscible polymers, if they are mechanically incompatible due to large differences in melt viscosities at the conditions of compounding. A simple approach of obtaining quantitative estimation of compounding conditions or grade selection of blend components through MFI has been suggested.[83] A methodology was developed for specifying the temperature and shear rate conditions for melt-blending two component polymers whose grades have already been selected based on other considerations. The temperature of compounding in order to achieve maximum mechanical compatibility is as follows:

$$T_c = \frac{E_{P_2} - E_{P_1}}{R \ln (\mathrm{MFI}_{P_2,T_{P2}} / \mathrm{MFI}_{P_1,T_{P1}} + E_{P_2}/T_{P_2} - E_{P_1}/T_{P_1})} \tag{4.17}$$

where $MFI_{P_1,T_{P_1}}$ and $MFI_{P_2,T_{P_2}}$ represent the MFI values of the two component polymers, T_{P_1} and T_{P_2} are the respective temperatures of MFI measurement, and E_{P_1} and E_{P_2} are their respective activation energies, which are given in Table 4.3 for a selected list of thermoplastics. They have been determined using the suggested method[75] based on MFI.

The compounding temperature calculated through Eq. (4.17) represents graphically the intersection point of the two curves shown in Figure 4.6 for polymers P_1 and P_2. Thus, if selection of the grades of the two component polymers has been done a priori, the compounding temperature as determined from the intersection of the two curves is automatically fixed. The value of T_c as obtained through Eq. (4.17) may not always result in a meaningful value for the selected grades. For example, the intersection points in Figures 4.6b, 4.6c, and 4.6d could be respectively too low (even below the melting temperature), too high (even above the degradation temperature), or more at all if the activation energies of the two components are equal. When such is the case, the value of T_c does not have actual relevance for use but is certainly an indication of the incompatibility of the component polymers.

The method for determining the shear rate level in a compounding operation for achieving mechanical compatibility has also been given through a simple equation, as follows:

$$\dot{\gamma}_c = MFI_{B,T_c}\left(\frac{K_{P_1}}{K_{P_2}}\right)^{1/(n_{P_2}-n_{P_1})} \tag{4.18}$$

where $MFI_{B,Tc}$ is the melt flow index value of either component, i.e., of the blend at the temperature of compounding T_c, K_{P_1} and K_{P_2} as well as n_{P_1} and n_{P_2} correspond to the parameters of Eq. (4.13) and those given in Table 4.6 for the two component polymers. The determination of the above shear rate condition is based on the intersection of the master curves using Eq. (4.16).

Figure 4.7 shows that when $n_{P_1} = n_{P_2}$, the curves do not intersect and hence no shear rate condition can be specified for achieving mechanical compatibility. When the K and n values for the component polymers are the same (as in the case of PET and PBT, where $K = 5 \times 10^4$ (g/cm · s^{2-n})·(g/10 min)^n and $n = 0.43$ for both polymers), then the blending shear rate truly does not matter because the two components are undoubtedly compatible, provided they have the same MFI value at the blending temperature. A few case studies are presented below to illustrate the above-postulated method of determining the compounding conditions (temperature and shear rate) and, alternatively, to select the appropriate grades of the component polymers if the compounding process parameters are fixed a priori. Note that the cases do not all involve engineering thermoplastics, which is the main focus of this book; however, they have been included here in order to exemplify the use of the above-discussed method.

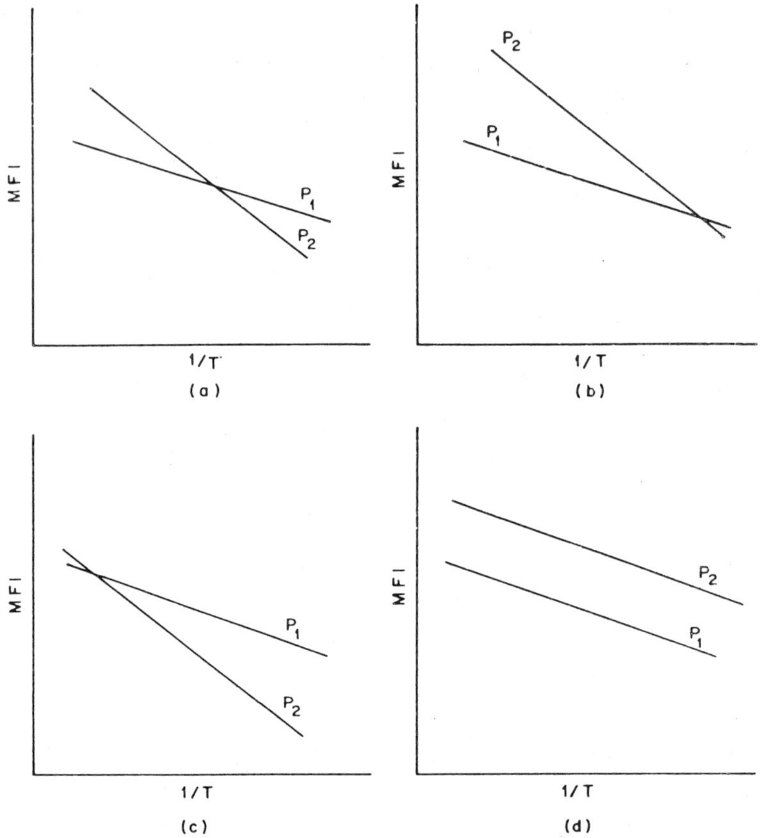

FIGURE 4.6 Four possible ways in which curves of melt flow index versus reciprocal temperature on a semilogarithmic plot would exist for two given component polymers. (From Ref. 83.)

Case Studies

Case 1: To determine the temperature and shear rate conditions for SAN/acrylic blend, we consider the same grade of SAN with three different grades of PMMA.

(a) Component 1: Tyril 860B [MFI = 9.5 (230°C/3.8 kg)]; component 2: Lucite 140 [MFI = 5.0 (230°C/3.8 kg)]. The relevant parameters needed for the calculations are given in Table 4.8. The loading and temperature conditions of MFI values for the two components are identical. Thus $T_1 = T_2 = 503°$ K. Substituting the appropriate values in Eq. (4.17) gives $T_c = 548°$ K = 275°C.

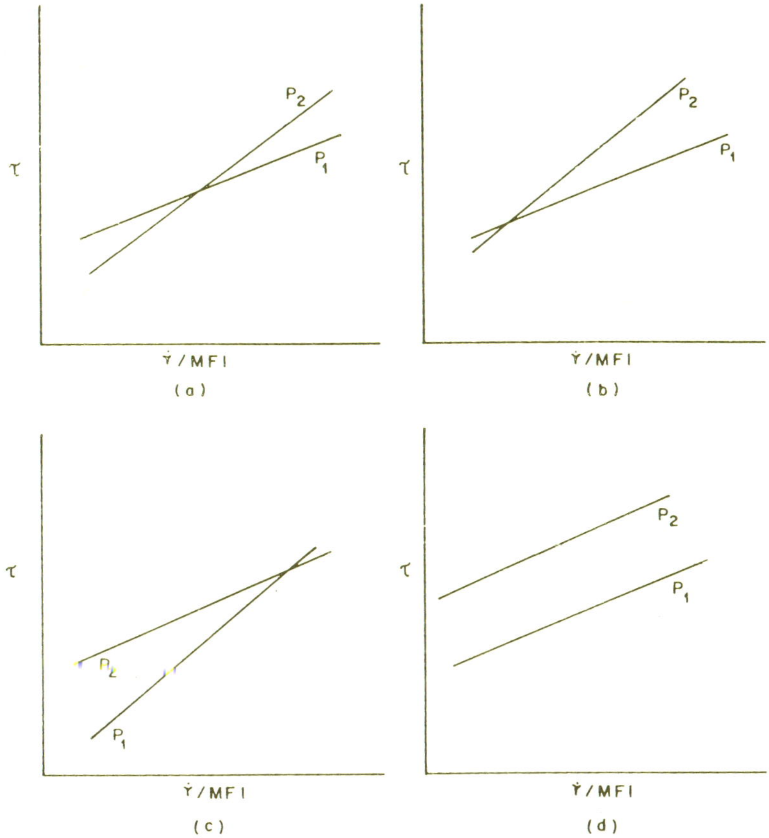

FIGURE 4.7 Four possible ways in which curves of shear stress versus shear rate normalized by MFI on a log-long plot would exist for two given component polymers. (From Ref. 83.)

(b) Component 1: Tyril 860B [MFI = 9.5 230°C/3.8 kg)]; component 2: Plexiglas VM100 [MFI = 7.9 (230°C/3.8 kg)]. Here again, $T_1 = T_2 = 503°$ K and hence, from Eq. (4.17), $T_c = 515°$ K = 242°C.

(c) Component 1: Tyril 860B [MFI = 9.5 (230°C/3.8 kg)]; component 2: Plexiglass VS100 [MFI = 11.2 (210°C/3.8 kg)]. Here $T_1 = 483°$ K and $T_2 = 503°$K and thus, from Eq. (4.17), $T_c = 410°$ K = 137°C.

It can be seen that in (a) and (c) the blending temperatures are a little high and too low, respectively. The better acrylic grade to blend with Tyril 860B is thus Plexiglas VM100 at a blending temperature $T_c = 242°$C. The

TABLE 4.8 Relevant Parameters for SAN and PMMA in the Case Study

Polymer type	Grade	MFI (temp °C/load, kg)	E (kcal/mol)	Applicability range of temp. (°C)	Load (kg)	n	K (gm/cm · s^{2-n}) × (g/10 min)^n	Applicability range of modified shear rate (s^{-1}) × (g/10 min)$^{-1}$
SAN	Tyril 860B	9.5(230/3.8)	30.7	200–250	3.8	0.33	3.0×10^5	3–1000
PMMA	Lucite 140	5.0(230/3.8)	38.5	170–260	3.8	0.44	3.2×10^5	100–1000
	Plexiglas VM100	7.9(230/3.8)	38.5	170–260	3.8	0.44	3.2×10^5	100–1000
	Plexiglas VS100	11.2(230/3.8)	38.5	170–260	3.8	0.44	3.2×10^5	100–1000

Table 4.9 Information on LLDPE and LDPE Grades for Blending

(a) Various LLDPE and LDPE Grades

Sample code	MFI (190°C/2.16 kg)
LLDPE A	1.0
LLDPE B	2.0
LDPE C	2.0
LDPE D	2.5
LDPE G	0.4

Source: Ref. 84. (Reprinted with kind permission of Society of Plastics Engineers, Inc.)

(b) Relevant Parameters for LLDPE and LDPE

Polymer	E (kcal/mol)	Applicability range of temp. (°C)	Load (kg)	$\eta_0 \cdot$ MFI	$\lambda \cdot$ MFI	N	Applicability range of modified shear rate $(s^{-1}) \times (g/10\ min)^{-1}$
LLDPE	3.2	175–205	2.16	9.6×10^4	6.0	0.11	0.01–2
LDPE	7.25	175–205	2.16	3.0×10^5	25.3	0.17	0.1–1

Source: Ref. 67. (Reprinted with kind permission of Elsevier Science Ltd., Kidlington, United Kingdom.)

MFI value of these two polymers at 242°C can be calculated from Eq. (4.10). In the present case of blending Tyril 860B with Plexiglas VM100, we get $MFI_{B,Tc}$ = 19.4 (230°C/3.8 kg). Using Eq. (4.18), the shear rate for blending can be calculated as equal to 10.8 s^{-1}.

Case 2: It has been shown that better blown films can be designed by using blends of LLDPE and LDPE.[84] The conditions of blending have not been given, but we illustrate in the following how easily this could be done by the above technique. The various grades of LLDPE and LDPE used are given in Table 4.9(a). The other parameters useful for performing the necessary calculations are given in Table 4.9(b). Various combinations of LLDPE and LDPE grades have been considered, and in each case the blending temperature has been determined as above. The results are given in Table 4.10. It can be seen that LDPE G requires a very high temperature for blending if it has to be mechanically compatible, thus showing that it is probably not the right grade choice for synergistic properties. The other component combinations would be mechanically compatible at different temperatures, hence a uniform blending temperature for all combinations of LLDPE and LDPE grades would not result in optimum property products.

In order to illustrate the method for determining the shear rate conditions for blending, we choose LLDPE A and LDPE C. Instead of the power-law model given in Eq. (4.16), we choose the modified Carreau model as given by Eq. (4.11), whose parametric values are tabulated in Table 4.9b.[67] Thus, comparing the shear stresses for the two polyethylenes using Eqs. (4.11) and (4.15) and solving,

$$\frac{(\eta_0 \bullet \mathrm{MFI})_{P1}\left(\dfrac{\dot{\gamma}_c}{\mathrm{MFI}_{B,T_c}}\right)}{\left[1+(\lambda \bullet \mathrm{MFI})^2_{P1}\left(\dfrac{\dot{\gamma}_c}{\mathrm{MFI}_{B,T_c}}\right)^2\right]^{N_{P1}}} = \frac{(\eta_0 \bullet \mathrm{MFI})_{P2}\left(\dfrac{\dot{\gamma}_c}{\mathrm{MFI}_{B,T_c}}\right)}{\left[1+(\lambda \bullet \mathrm{MFI})^2_{P2}\left(\dfrac{\dot{\gamma}_c}{\mathrm{MFI}_{B,T_c}}\right)^2\right]^{N_{P2}}} \quad (4.19)$$

The $(\dot{\gamma}_c/MFI_{B,Tc})$ value can be determined, which in the present case is found to be equal to 37.6. Now $MFI_{B,Tc}$ can be calculated using Eq. (4.10) to give a value equal to 0.578, and hence the shear rate for blending is 21.7 s^{-1}.

Case 3: Suppose we decide to make blends of LLDPE and LDPE. Then the choice of grades can be made as follows based on the predetermined blending temperature.

Using Eq. (4.17), the ratio of $MFI_{P_2,T}$ to $MFI_{P_1,T}$ can be calculated for different blend temperatures. Thus, for LLDPE and LDPE blends, the results obtained are as shown in Table 4.11. It can be seen that based on the choice of

TABLE 4.10 Compounding Temperature for Various LLDPE/LDPE Blends

Component 1/Component 2	Compounding temperature (°C) calculated from Eq. (4.17)
LLDPE A/LDPE C	127
LLDPE A/LDPE D	110
LLDPE A/LDPE G	312
LLDPE B/LDPE C	190
LLDPE B/LDPE D	168
LLDPE B/LDPE G	457

the blending temperature and the grade of one of the components, selection of the grade of the other component can be made.

A number of such cases can be cited wherein the above method can be used effectively for selecting the grades of polymer components for the blends or, alternatively, the blending conditions. This technique is the simplest and quickest route for determining compounding conditions for mechanical compatibility.

In the present chapter, the major emphasis was on shear viscosity and its potential for use in processing based on the simple idea of utilizing MFI rather than any other measure of viscosity. However, it is known that the deformations in any polymer process are of both shear and extensional types. It is not just the shear viscosity that is important, but in certain situations the elasticity of the polymer plays a dominant role, whereas in other cases it is the extensional viscosity that takes the frontline. Estimates of the variation of elasticity and extensional viscosity with shear rate then become necessary if calculations are to

TABLE 4.11 Required MFI Value Ratio for LLDPE/LDPE Blends at Various Compounding Temperatures

${}^{MFI}LLDPE,T/{}^{MFI}LDPE,T$	Compounding temperature T_c (°C)
2.0	125
1.5	150
1.16	175
1.0	190

be made with respect to the processing parameters of relevance. A method for obtaining unified curves of normal stress difference versus shear rate[85–87] as well as extensional viscosity versus shear rate is available.[88] It has also been shown that dynamic viscoelastic data can also be coalesced in a similar manner.[89] In processing operations, where these rheological parameters achieve significance, an approach similar to that discussed in this chapter can be followed in order to obtain estimates of processing parameters that are affected by changes in elasticity.

5

The Injection-Molding Process

PROCESS DESCRIPTION

Since most engineering plastics are consumed in the injection-molding process, it would be helpful for those concerned with selecting the best candidate resins for their applications to understand the basics of the process. It is not the intent here to prepare an operating manual for potential molders, but rather to provide a background for later discussions involving the processibility considerations and part costs that arise from the process. Those interested in pursuing the molding process in depth should refer to the excellent and thorough presentation in *Injection Molding: Theory and Practice* by Irving I. Rubin.[90] Other books on polymer processing also deal with injection molding and hence may be referred to for better understanding.[91–97]

The theory of the operation of a reciprocating injection-molding machine is simple. The machine (see Figure 5.1) is provided with equipment to feed a plastic resin in the form of small cubes or granules to a hold-up bin or hopper, which provides a flow into a heating cylinder where the resin is melted and then forced under high pressure into a mold clamped closed under high pressure. The melt solidifies and cools in the mold, after which the mold is opened and the part removed; the mold is then closed, and the whole process is repeated. It is an intermittent, reciprocating operation.

SCREW INJECTION MACHINE

Hopper

The resin hopper of a molding machine is usually sized to hold at least several hours of feed. It is located at the back (or feed) end of the machine, and has

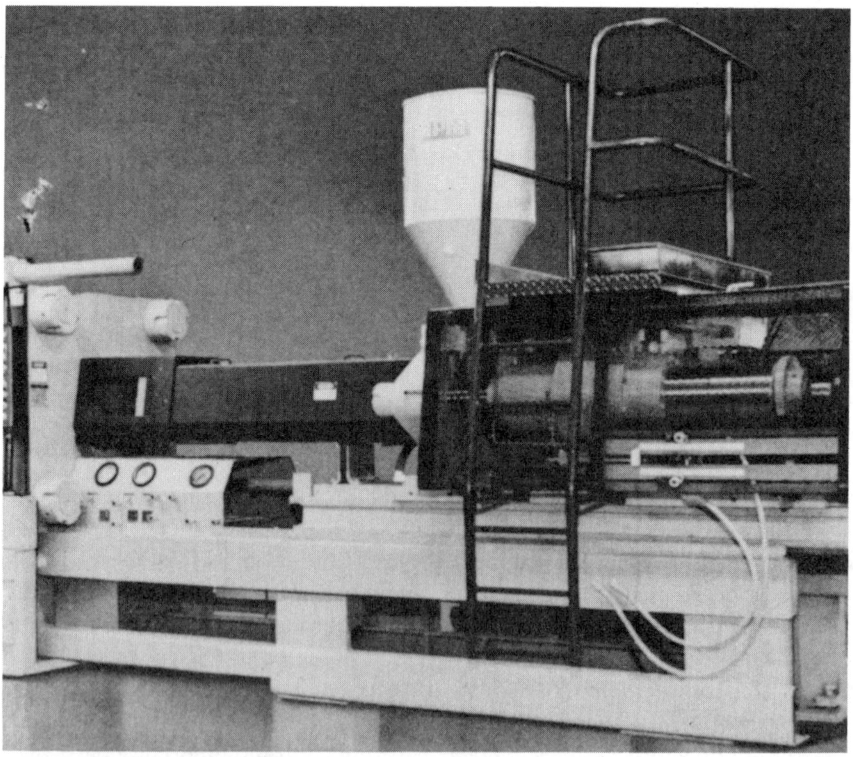

FIGURE 5.1 A typical screw injection molding machine. (Courtesy of HPM Corporation, Mount Gilead, Ohio.)

sloped sides to encourage gravity flow of the resin pellets into the feed throat of the extruder barrel.

Hopper Dryer

A significant number of engineering plastics, such as nylons, ABS, polycarbonates, and polyesters, absorb moisture from air and need to be dried before molding if they are to perform to their capabilities. If the moisture is not removed before processing, it will accelerate equipment corrosion and cause problems in the forming stage.

The energy used for drying can be significant; at times it can be of the same order, per pound of resin, as the total of all other energy used in processing.[98] Part of the drying energy is lost in water vapor, and part goes to heat up the resin. One approach is to vent the plasticating device to allow the outflow of water vapor and thus eliminate a separate drying step. Another approach is to use a hopper dryer so that the resin enters the plasticating machinery immediately after drying and carries the drying heat with it. Most modern molding shops provide their machines with desiccant hopper dryers (see Figure 5.1).

The desiccant hopper dryer system is a closed system, with blowers forcing hot dry air up through the hopper to remove the moisture from resin. The moisture-laden air is directed through a filter to a desiccant cartridge, usually filled with super-dry zeolite as a drying agent. This dry air is again recycled through the heater and then to the hopper.

The temperature of the hopper dryer can be varied with the resin to be dried, but an average temperature is about 175°F. It is important to the molder to insulate the hopper dryer to maintain the temperature of the hot air entering the hopper. As a rule of thumb, the air flow should be numerically the same as the number of pounds per hour of resin to be dried; that is, for 60 lb/h of feed to the molding machine, the volume air flow rate to the hopper should be 60 ft/min.

Normally, material suppliers will provide suggested drying times and temperature for their resins, keeping in mind on the one hand that the resin as shipped is usually dry, and on the other that a certain amount of regrind or rework containing high moisture levels will frequently be a part of the feed to the hopper. The molder should follow the hopper dryer suggestions of the suppliers because, for example, 275°F is appropriate for a polyester, but a nylon would begin to darken in color if dried at that temperature. The nylon would prefer a drying temperature of about 175°F.

Perhaps this is too much detail about drying, but resin selectors must understand its importance and the consequences if their molding candidates are wet during molding.

Barrel and Screw

The purpose of the extruder barrel and screw as a unit is to convert the solid plastic pellet fed from the hopper into a homogenous melt that can be forced under high pressure into a closed mold.

The screw is typically machined into three zones (Figure 5.2): a feed zone, usually 50% of the length of the screw; a transition zone, where most of the actual working and melting of the plastic takes place, usually 25% of the screw length; and a metering or pumping zone covering the remaining 25% of the length. The channel depth of the feed zone is the deepest of the three. In the transition zone, the channel depth decreases until it ends at the metering zone, where the channel depth of the screw is the shallowest, normally 30–40% as deep as the feed zone.

The depth of the screw channels is an important consideration with respect to the output rate and melt quality. The most effect of flight depth on output is that the deeper the channel, the more rapid is the decrease in output with increasing back-pressure. A shallow screw (particularly the metering section), while it has a much lower output, is relatively insensitive to the back-pressure setting of the molding machine. The molder must be familiar with the relationship between the melt viscosity of the particular engineering plastic being considered for an application and the efficient depth of the screw channels. A method is available for selecting the "effective viscosity" for isothermal flow of polymer melts in screw pumps.[99] Curves are also available for calculating the efficiency of the screw channel, which can in turn be used for estimating the energy dissipated in screw pumps.[99] Basically, if the channels are deep and the melt has a low viscosity, unmelted particles in the melt are possible. Furthermore, output pressure of the melt is decreased and nonuniform. Conversely,

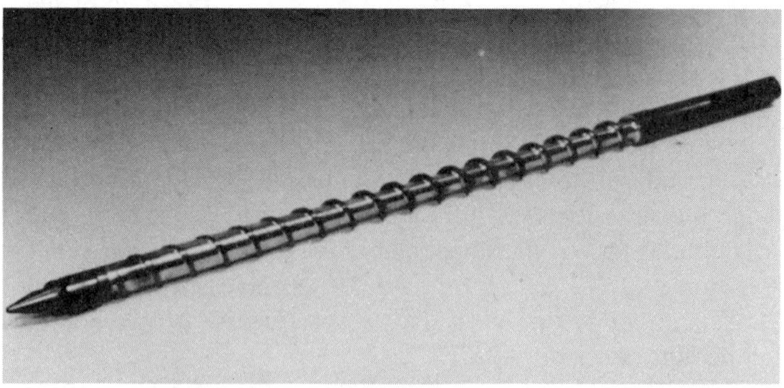

FIGURE 5.2 A typical three-zone screw for an injection-molding machine. (Courtesy of E. I. du Pont de Nemours and Company, Wilmington, DE.)

if the channels are too shallow, the resulting shear rate is likely to result in resin degradation.[90]

Again, it is not really important that resin selectors know these facts intimately. What *is* important is that they understand what the molder may say with regard to their screw design and how it might affect their opinion of which resin will run best in their equipment.

The barrel for a reciprocating screw machine must be able to contain at least 20,000 psi of injection pressure, which can be generated during operation of the machine. It is equipped with external, instrument-controlled electric heaters. The diameter of the hole in the barrel is the nominal diameter of the screw, with very little clearance.

Although the barrel heaters provide some heat input for melting, and can be used to exercise some temperature control on the process, the turning of the screw (usually by hydraulic drive) converts mechanical energy into enough heat to melt most of the resin. As the material melts, it moves forward along the screw flights to the front end of the screw. The pressure generated by the screw on the material forces the screw, the screw drive system, and the hydraulic motor back, leaving a reservoir of plasticized material in front of the screw. Pressure generated by the hydraulic motor is then diverted to force the screw forward, injecting the material into the mold cavity.[90]

Nozzle

It is through the nozzle that the polymer melt is injected into the mold. In the case of multistation systems, the shape and outside dimensions of the nozzle have to be such that it mates positively with each mold. With an increase in nozzle diameter, there is an effective decrease in the pressure drop, injection time, and temperature rise. The nozzle usually has a reverse taper of 4–5° and as short a land as possible. The nozzle bushing on the mold also has reverse taper and an inlet bore which is the same as or slightly smaller than the nozzle.

Runners

Runners are channels provided to convey the polymer melt from the injection point, namely, the sprue, to the mold cavities. They are usually trapezoidal, half-round, or round in cross section. They are short and direct, streamlined, and slightly tapering to reduce pressure loss. Basically, it is the thickness of the molding at the gating point that governs the cross-section size of the runner. For multicavity molds, it is important that the lengths of runners between the sprue and cavity be the same so that all cavities are filled at the same rate under the same effective pressure. This requirement becomes essential when the product has to meet close tolerances. Whenever the runner system is extensive, it is advisable to provide the runners with vents in order to help reduce the volume of air and gases which have to be removed via the cavity vents and flashlines.

Gates

Gates are restrictive passageways from the runner to the cavity and are sized to allow easy flow of the polymer melt into the cavity. Generally, they are full-round, fanned, with a bore one-half of that of the feed runner. They are kept as short as possible. The final sizing of the gates is usually done by carrying out a practical molding test. Actually, it is the gate position rather than the size which is more important. Normally, it should be positioned in such a way that it feeds the thickest section of the molding in a nonstressed area of the molding. The gates position truly becomes critical whenever weld lines are likely to be a problem.

Mold

The injection side of the mold is clamped to a stationary platen, and the ejection side of the mold is clamped to a moving platen. The mold has an empty space in the configuration of the part to be molded. This empty space is what is filled by the melt under high pressure from the screw. The mold is held closed during this injection step by hydraulic pressure applied to the movable platen that is sufficient to overcome the injection pressure in the mold.

The injection molder and the molder designer know the importance of operating under minimum pressure drop conditions during cavity filling in a processing operation in order to minimize frozen-in stresses and the clamping forces preventing mold opening. The estimation of the pressure and clamping requirements for mold filling is important to the injection molder and especially to the mold designer when there is a desire to change the type of polymer being processed or when there is a change of grade of a polymer of the same generic type.

A simplistic approach has been suggested for estimating the minimum pressure drop during mold cavity filling based on an order-of-magnitude approximation technique.[78] The expression for the minimum pressure gradient $P_{0,\min}$ for center-gated disk mold cavity filling is given as:[78]

$$P_{0,\min} = \frac{f(n) K \overline{C}^{3n} \pi^n R_d^{1+n}}{MFI^n n^{1+4n}} \tag{5.1}$$

where

$$f(n) = \frac{2^{3n+1}}{1-n} \left(\frac{1+5n}{3n} \right)^{3n} \left(\frac{1+5n}{1+2n} \right)^{1+2n} \tag{5.2}$$

K and n (The model parameters given in Eqs. (4.13) and (4.16), whose values are tabulated in Table 4.6. MFI is the melt flow index value at the temperature of interest. x_0 is the original cavity thickness, R_d the radius of the circular disc cavity, and $\bar{C}$ a proportionality constant, given as

$$\bar{C} = 2a_H^{1/2}\left(\frac{T_0 - \theta_m}{T - \theta_m}\right) \tag{5.3}$$

a_H being the heat diffusion coefficient of the melt, θ_m the mold temperature, T the melt temperature, and T_0 the freeze-off temperature.

For a given polymer type, $f(n)$, $\bar{C}$, K, and n are all constants, and for a given circular type, mold R_d and x are fixed. Hence, the minimum pressure for cavity filling can be estimated from Eq. (5.1) through knowledge of the MFI of the polymer at the melt temperature T of mold filling. The MFI determined under standard ASTM test conditions would have to be converted to MFI value at the required temperatures by using either Eq. (4.9) or (4.10).

Under Eq. (5.1), the minimum clamping force F_{min} can be easily estimated from the following expression:

$$F_{\text{min}} = \frac{P_{0.\text{min}}\bar{A}}{\bar{B}} \tag{5.4}$$

where $\bar{A}$ is the effective projected area of the molding, given as πR^2, and $\bar{B}$ is a numerical function which depends on n as shown for circular, square, and rectangular moldings:[100]

$$\bar{B} = \frac{n+2}{n} \quad \text{(for circular panel with center injection)} \tag{5.5}$$

$$\bar{B} = \frac{n+1}{n} \quad \text{(for long, thin rectangular panel with center injection)} \tag{5.6}$$

Thus more centrally gated, flat, and symmetrical moldings would show higher values of $\bar{B}$ and hence, a lower clamping force. For engineering design calculations, one could chose Eq. (5.6), which would give the lower estimate of $\bar{B}$. The A value of $\bar{B} = 4.0$ for $n = 0.33$. $\bar{B} = 3.5$ for $n = 0.44$, and $B = 3.0$ for $n \approx 0.50$ could be used as rough estimates, when the n values of the considered polymers are close to these values.

When the melt has been properly injected into the mold, usually provided with cooling equipment, and the melt has attained the solid state, the mold is opened and the part is removed.[90]

The molding cycle may be defined as the time interval from one completed shot to the next. The inexperienced resin selector must understand that the molding cycle, which is a critical factor in part cost, usually varies from one polymer to another, and with melting rate, cooling rate, part thickness, and other variables. Molders are the best qualified to estimate what the molding cycle will be for a given resin (provided they have run it in other parts) and a given part, and this information must be included in the cost-estimating procedure covered in Chapter 9. If the molder is not familiar with a particular resin, the material supplier must be contacted for an estimated molding cycle for a given part thickness, etc.

6

Kinds of Injection Molders

THE CAPTIVE MOLDER

It has been estimated that 30% of the plastics used in engineering applications are converted by captive molders. These usually involve large companies that have conceived the need for the plastic part, designed it and the mold required to produce it, and produced it in-house in large commercial quantities. A major automotive manufacturer is a good example.

A captive molding shop, itself, is usually large, although there are exceptions. Whenever possible, molds are designed for unattended machine operations, and every attempt is made to make them idiot-proof. In a way, this parallels assembly-line conditions, where shift operation is the rule, and overseers for the machines are subject to transfer to other duties as they are needed. When such molding lines are operating correctly, which is most of the time, there is little need for the constant presence of a skilled molding machine operator. Such mass production can and should be the most economical method of producing large volumes of good products.

It depends on methods of accounting, but usually the captive shop is not dedicated to profit making per se. Rather, its mission is to produce quality parts at the lowest reasonable cost, so that they, as components of a larger unit, will result in a profitable final product.

In terms of resin selection, managers of a captive molding shop play a fairly small role, usually as part of a team that includes designers and material engineers of their firm.

THE CUSTOM MOLDER

Custom molders, unlike captive molders, are independent businesspersons, and must make a profit in their operation to survive. Further this influence on resin selection is significant, for a number of reasons to be discussed later.

The custom molding shop is usually smaller than a captive shop, but can vary from one with only one machine to one having a hundred or more. Both of these are extremes of an average or "typical" shop. Actually, it is difficult to arrive at a valid description of an average custom molder, perhaps because there are so many shops and because they are so uniquely set up as a result of rapid growth over the past 30 years, with many different solutions to a number of common problems. In any case, then, instead of an average molder, a "typical" shop will be described in order to simplify the understanding of a custom molding operation by the resin specifier.

Typical Custom Molder

Many factors need to be considered before selecting the number of machines of auxiliary equipment needed to define a "typical" operation, but the example presented here will provide end users with a grasp of the size and extent of equipment usually available to them from many molders.

Perhaps the optimum number of molding machines for a profitable small custom injection molder ranges from four to eight. Fewer machines generally have a high initial investment per machine, while more than eight generally (1) require disproportionately more operating personnel, (2) cause the overall operating utility of the shop to go down, and (3) make the operation more difficult to manage efficiently. Of course, there are larger shops than this that *are* run effectively, but it presents a greater challenge. For the example here, an eight-machine molding shop has been used.

The size of the machines will depend on the ongoing commitments for orders and on the planned future of the business, but usually a range of machine sizes is appropriate.

A useful machine size range would be from 3–5 oz to 20–40 oz, where the ounce designation refers to the nominal shot weight capacity of a given machine. For this example shop, there would be three 3–5 oz machines, three 6–12 oz machines, and two 20–40 oz machines. This mix would provide a good degree of flexibility to molders, permitting them the opportunity to bid on a high percentage of the jobs available in the marketplace.

Of course, the other way to rate molding machine output is by clamp capacity, which, along with the screw and barrel size, controls the output. Combining the two variables that determine production rate, the annual capacity of the eight machines selected to represent a typical shop would look like what is shown in Table 6.1.

The figures in Table 6.1 are estimates, and they take many factors into consideration. The output of a machine depends greatly on part size, run length, mechanical utility, clean-up time between runs, and others. The figures shown are estimates that assume: (1) a cycle evenly split between injection time and hold time, (2) 24-h operation, 5 days/week, 50 weeks/year, and (3) an 80% overall operating utility. Ranges for each size machine are due to variations among manufacturers in throughout rates. These outputs allow for a reasonable number of bad parts that must be reworked.

The investment in these machines, plus essential auxiliary equipment and building, is significant. It is an important aspect of the resin selection process that the person or company wanting an engineering part made by a custom molder appreciate the extent of this investment and the resulting effect on part costs. The itemized equipment costs, shown in Table 6.2, are estimates by the authors, simply to illustrate the high capital requirements of a reasonably sized shop, and are not to be used by anyone interested in setting up a new shop.

Cost of Manufacture (Excluding Materials)

Without getting too involved in the details, an assumption is made here that a force of 22 people, excluding management, is a reasonable size to operate our

TABLE 6.1 Machine Capacities of Typical Molding Shop

No. of machines	Rated capacity (oz)	Clamp pressure (tons)	Estimated annual output (K lb)	Conservative average, annual output (K lb)	Total annual output (K lb)
3	3–5	75	180–360	200	600
3	6–12	200	480–720	500	1500
2	20–40	400	1080–1560	1000	2000
Total					4100

TABLE 6.2 Total Permanent Investment—Typical Custom Molder

Equipment investment		Cost (K dollars)
Molding machines		
Three 3–5 oz (75–ton clamp), installed		200
Three 6–12 oz (200–ton clamp), installed		350
Two 20–40 oz (400–ton clamp), installed		350
Drying oven		10
Six scrap cutters, installed		40
Eight water temperature controllers		30
Mold handling facilities		40
Eight hopper dryers		50
Auxiliary mold equipment		50
Miscellaneous instruments and controllers		5
Miscellaneous materials handling equipment		15
Pollution control equipment		5
Machine shop equipment for mold maintenance (includes lathes, milling machines, welders, grinder)		200
		1345
Space and utility investment (ft^2)		
Operating space for eight machines	2,700	
Mold maintenance area	150	
Mold storage area	200	
Resin storage area	2,000	
Part storage, packaging, shipping	8,000	
Offices, lunchrooms, rest rooms, etc.	1,500	
Unloading dock	100	
	14,650 @ $25/$ft^2$	375
Utility requirements		10
Total permanent investment		1730

"typical" shop, assuming it will operate three shifts per day, 5 days per week. Applying average estimated 1980 wages, salaries, and benefits for the mix of the 22 employees involved, $25/h machine costs were determined as an average rate for the three different size machines given in the example. Individually, the labor cost for the 3–5 oz machines is estimated to be $20/h, for the 6–12 oz machines $25/h, and for the 12–20 oz machines $30/h. These are in line with the rates used in the "cost estimator form" issued by the du Pont Company as an aid to its customers and the end users. This form to be covered in detail in Chapter 9, is a major tool in the resin selection process.

Expertise of the Custom Molder

It is necessary to point out that there is more to an efficient molding shop than new equipment and plant facilities. A major requirement is that management of the shop understand the molding process thoroughly. If there were such a thing as a priority list of work skills the management should have, it might be in this order:

1. Mold design and operation
2. Machine operation
3. Costs
4. Sales
5. Administrative

The management of a custom molding shop needs a number of other skills, of course, but end users selecting molders and relying to some extent on their judgment in making a selection of resin should examine carefully the molders' competence in these five areas of expertise.

Mold Design and Operation

The primary function of a mold is to shape the finished product. In order to do this, it must have some means of introducing the plastics material to be formed; it must have some means of forming the inside, outside, and both sides of the product; it must have some means of maintaining the temperature desired in the process; it must have some external or internal mechanism for operating the various features of the mold; it must have some means for allowing the finished product to be removed from the sections which were used for the forming process (this is called ejection); after ejection, it must provide for easy removal of any and all excess material that may have been left during the previous cycle; it must be designed and constructed with adequate strength in the various sections to resist the alternate application and release of pressure.[101]

A list of general mold design requirements is given below.

1. The mold should be rigid enough to avoid deflection under pressure, which can cause flashing.
2. The guide dowels and bushes which are used in lining up the mold halves should mate precisely and lock rigidly. Further, they should be sized to allow for thermal expansion.
3. The working faces of the mold, especially on the split line or closure faces, should be kept clear from screw fixing holes and joints.
4. All surfaces coming in direct contact with the polymer melt should be highly polished to ensure good surface finish on the molded product.

5. Shrinkage allowances must be made in the cavities and on the core pines, depending on the selected polymer to be processed.
6. When designing the mold, due consideration should be given to the transfer of the mold into and out of the machine. Thus, suitable tappings and holes should be provided for the attachment of lifting devices.
7. Shallow grooves should be provided from the cavities to the atmosphere on the mold split line and on the opposite side of the gate or feed points to allow for escape of air and volatiles.
8. Trim grooves should be provided around the cavities to give clean moldings and minimize extra after-trimming operations.
9. Land faces should be provided around feed runner, gate, and cavity areas in order to achieve a good bite between the split line faces of the mold and thereby minimizing flashing when clamping force is applied.

The above introduction to this section on mold design and operation covers the basic function of the mold and its design features. it is important to understand mold function and design, because the mold is the heart of the injection-molding process. It contains the cavity (or cavities) that receive the hot melt under great pressure (up to 20,000 psi) from the injection end of the molding machine. When the part is cooled appropriately and removed from the mold, it must confirm to all the specifications imposed by the end user. It is obvious, then, that the quality of the part is largely dependent on a mold that is well designed and well made.

A good mold design can be done based on flow simulation using a system approach.[102] By simulating the process with different sets of mold design parameters and process control conditions, a list of alternative mold designs and associated molding conditions can be obtained for a given material. The output [102] from the simulation also includes such information as predicted full time, shrinkage, anisotropy, or other characteristics of interest which would be useful to the mold designer in making a final decision. The simulation can be repeated for different materials and hence can be used in aiding the selection process for getting the appropriate choice of material.

With regard to the mold design, it is essential that the mold designer work closely at the very beginning with the part designer. Obviously, if the part is improperly designed, it will not be acceptable; but it is unusual to make an acceptable part from an improperly designed mold. Therefore, early in the resin selection process, the part designer and the mold designer must agree that the polymer selected has processibility characteristics that are well understood by both of them.

This point is emphasized here because at times the custom molder also fabricates the mold. Normally, however, the molds are made by toolmakers. In any case, in the design stages of the mold, the end user, the molder, and the toolmaker should all be consulted before cutting steel. The critical consideration in mold design is confidence that it will work successfully, for it is much less costly to change mold design than it is to recut a heavy piece of steel.

Many considerations go into mold design and fabrication. The resin specifier should have an appreciation of a rather lengthy list of important factors, as shown in Table 6.3.

To repeat, injection molders must be completely familiar with all of these criteria, not only to apply the knowledge to their operation, but also to advise the end user early in the resin selection and mold design process. In turn, it is advantageous to the end user and part de-

Table 6.3 Mold Design and Fabrication Considerations

Number of cavities	Runner system
Material	Hot runner
Steel	Insulated runner
Stainless steel	Gating
Prehardened steel	Edge
Hardened steel	Restricted (pin point)
Beryllium copper	Submarine
Chrome plated	Sprue
Aluminum	Ring
Epoxy steel	Diaphram
Parting line	Tab
Regular	Flash
Irregular	Fan
Two-plate mold	Multiple
Three-plate mold	Ejection
Method of manufacture	Knock-out pins
Machined	Stripper rings
Hobbed	Stripper plate
Cast	Unscrewing
Pressure cast	Cam
Electroplated	Removable insert
EDM (spark erosion)	Hydraulic core pull
	Pneumatic core pull

signer to have a working knowledge of them in order to consider the advice intelligently.

Machine Operation

As time passes and the plastics processing industries become progressively more sophisticated, it is logical to look forward to a time when the operation of most injection molding machines will become completely automatic. This will require special molds, such as hot runner molds, and computer feedback mechanisms for the machines, all of which are available to some extent today. But until such general automation becomes a *fait accompli*, owners and managers of a molding shop, as well as their production and maintenance crews, must be intimately familiar with the injection-molding machine and its operation.

Knowing the equipment is not enough by itself, however. It is also necessary for molders to have a basic, if only empirical, understanding of the rheological characteristics of the resins[61] that are to be reviewed during the resin selection process. Further, they must thoroughly appreciate which polymers must be dry before processing; how much regrind (%) in the feed to the machine is reasonable for each resin. They must appreciate the fact that different plastics react differently to the same heat history or, conversely, that each polymer will have its own hold-up time and temperature limits before degradation becomes a problem.

It is very important that injection molders be familiar with the cycles that can be obtained with the leading engineering plastics. They will know by experience how to *estimate* the optimum cycle for each polymer according to part thickness, which is a critical input in part cost estimating. Further, they will be familiar with operational adjustments to the equipment that will minimize the *actual* cycle obtained during the production run.

Costs

An obvious skill required by the custom molder is the establishment and maintenance of a realistic costing system in the shop. It must reflect accurately the cost of doing business, and a realistic appraisal of the profit and loss. Further, it must measure part costs accurately, since it will provide input into the estimated costs for new parts. Although a review in depth of costing techniques for custom molders is not within the scope of this discussion, resin specifiers must have confidence in the custom molder's ability to determine their costs accurately, for the resin selection process—or even the entire project—could fail if the initial estimated part costs are misleading.

Sales

The sales function of the custom injection molding shop has not much direct bearing on resin selection. It is necessary, however, for a prosperous, efficient shop to have at least one direct sales representative, or a good sales agent, bringing in potential and actual new business on a routine basis. The sales representative must have an in-depth understanding of part and mold design, and the ability to derive realistic part costs for bidding purposes. An end user dealing with a capable sales representative is most likely to be favorably impressed by the custom molder's operation, and in indirect ways, can be influenced in resin selection by the sales representative's preference for one resin over another, when properties of candidate polymers are comparable for the end-use application.

Administrative

There are no particular administrative skills peculiar to managers or owners of a custom molding shop that would not apply to any small business, but, without them, the dependable supply of parts to end users is always questionable. They should be able to see to it that safety, equipment maintenance, housekeeping, product quality, shipping schedules, and employees' morale are under control; and end users will then be favorably impressed with their dependability and their input to the resin selection process.

7

Roles of Custom Molders in Resin Selection

SERVICING LARGE ORGANIZATIONS

Many custom molders supply precision-molded engineering parts to large customers such as the automotive industry, perhaps as much as 70% of their requirements. Usually, the resins used are specified by the automotive company, which has already devoted considerable attention to the engineering aspects of the resin selection process.

In a discussion of the problems of servicing the automotive industry, one operations manager at a large custom molding shop said, "Resin selection is one of our biggest headaches. It can be an absolute mess, and we have lost out a number of times by picking the wrong resin."[103] Again, this was directed specifically to the automotive industry, but could really apply to any major company.

The resin selection process in the automotive industry is unusual, and can be confusing to everyone, including the specifier, the molder, and the material supplier. For example, one of the "Big Three" automotive companies, on a quote or an order, lists the material by its generic name and gives the code number of key specifications. The material (resin) can be looked up on separate materials lists, which give further details. This list includes all other materials in addition to plastics, such as paint, rubber, steel, and aluminum. Due to its sheer size, the list is often out of date, since it takes time to get new materials listed. Processors (molders) then have to locate the approved supplier on a separate listing of all approved suppliers for all kinds of materials, parts, and services. Again, problems of size tend to render that list incomplete.[103]

The custom molder must deal with this complexity, often without an understanding of why the particular resin was specified. Another problem is that many molders and subcontractors are not aware that these lists even exist, and just pick a resin based on its description by calling a supplier who may not be on the approved source list.[103]

The obvious answer to the problem of these inefficiencies in selecting resins on the part of the large companies, and to the problem of the custom molder misunderstanding the specifications and unilaterally selecting the wrong resin, is better communications. It is also, however, a matter of both the specifier and the molder understanding better how to select the appropriate resin based on logical technical considerations.

SERVICING SMALLER ORGANIZATIONS

Custom molders accept many jobs from smaller end users, who may not be very sophisticated in their approach to the resin selection process. In fact, some end users may never have used plastics in their products before. When this is the case, custom molders and end users commonly meet to consider the resin selection question, with end users providing the in-use requirements of the part to be molded and custom molders helping to develop the resin decision based on empirical judgment arising from their previous experiences. Frequently, the two will consult the materials suppliers before finalizing the decision; but there are occasions when this check is omitted.

The concern about this simple approach is the probability that important possible resin candidates are never even considered. It verges on the impossible for molders or end users to have reasonable access to literature on hundreds of resin compositions, and it verges on the unrealistic to expect material suppliers, when consulted, to refer the molder to a competitive resin that may be slightly more appropriate than their product. Nevertheless, although this simple procedure can be incomplete, it frequently works, for it is only occasionally that the resin selected by it fails in the application. On the other hand, there are times when it is inadequate or when overkill is involved, and the most economical choice has been overlooked. The point of this observation, however, is neither to criticize nor endorse this particular route to resin selection, but rather to bring its existence to the reader's attention. It is part of the real world, and includes judgment, some technical background, experience, and can even involve the consideration of personalities or personal preferences on the part of custom molders.

This personal preference, or pragmatism if you will, does enter the molders' selection process occasionally once they are confident that one or two of the polymers commercially available to them will provide all the properties needed by the application. Pragmatism in this context is meant to suggest what is best for custom molders after their customers' minimum requirements have been satisfied. Following are some of the decisions made by custom molders that are motivated by their desire to produce parts acceptable to their customer while making their efforts as simple and as profitable as possible.

Personalities (Salesmanship)

Not even in a technical treatise on resin selection should the consideration of personalities be omitted, particularly in view of the number of options custom molders may have. Material suppliers usually have competent, marketing-oriented technical sales people to call on molders to sell their products. Aside from the merits of their products, the personalities of these people are important. Molders who are comfortable with resin sales representatives will normally try to use their products. If a specific resin, available from a number of suppliers, is being considered, naturally the favorite sales representative will get the business. Beyond that, however, molders can, and sometimes do, ask sales representatives if they have a product equivalent to or better than the one being considered. The sales representative can then ethically switch the new application to a polymer that is not even in the same family as the one originally proposed. Thus, an "engineering" decision can be made based on personalities.

Material Supplier Performance

Performance is another consideration on the part of molders, but one that is appreciably more objective. Molders have the right to expect good service from their material suppliers. They want delivery within a reasonable time, and at a time that suppliers have agreed they can meet. Molders' profits are very seldom exorbitant, nor are they normally blessed with a surfeit of working capital. Consequently, they do not want to stock excessive inventory to guarantee supply. Thus, the supplier's delivery schedules must be dependable.

The average molder will prefer a quality resin. Again, the material supplier must be dependable, providing resin with uniform quality. Off-quality product obviously can be a very serious impediment to the molder's efficiency, and if the parts they produce are unacceptable to the end user, they must accept their return at an economic penalty. At the same time, the end user can be seriously inconvenienced.

If it happens that the resin quality is below standard, molders expect quick action on the part of suppliers. They expect suppliers to provide rapid evaluation of their quality complaints, and some consideration in defraying their production delays and cost penalties.

Most custom molders are capable of solving a number of molding problems as they occur from time to time. However, there are occasions when they cannot, and it is then that technical assistance is requested from the supplier. The assistance should be prompt and effective. Further, assistance in mold and part design is deeply appreciated by molders.

All these demands on material suppliers by molders have an influence on resin selection in one way or another. If their experience in the past with suppliers in any of these problems has been less than satisfactory, molders will try to influence end users on resin selection, not only between two suppliers offering essentially the same product, but even to the point of recommending a different polymer altogether.

Mold Shrinkage

The crystalline polymers, in most of their forms, have high shrinkages compared to those obtained with amorphous ones. Higher shrinkages normally present more difficulties in the design of a mold and, at times, in its operation. For one thing, the crystalline resins are usually anisotropic, and parts will shrink more in one direction of the molded part than it will in another. That is not to say that this problem cannot be solved by careful mold design considerations, but it is more exacting, and occasionally more expensive, than designing a mold for an amorphous resin. It must be kept in mind that, in spite of shrinkage considerations, a large proportion of engineering plastics are specified in crystalline polymers because of their superior properties.

The molder is concerned not only with shrinkage uniformity, but also with the extent of shrinkage, since cutting steel when high shrinkage problems are a known problem can be a nerve-racking experience to the mold designer, mold maker, and molder alike. If the specified dimensions are not obtained on the part after initial mold sizing, the mold must be modified until acceptable dimensions can be obtained with attendant delays and increased costs for all.

If, in good faith, molders feel that an amorphous resin can provide parts with adequate properties for an application, even though the properties of a crystalline polymer would be better, they will recommend it to the end user. On the other hand, there are occasions when part design calls for little or no draft in a major surface area; and the amorphous (low shrink) polymer can only be removed from the mold with great difficulty, in which case the molder will lean toward the use of a crystalline polymer when a choice is available. Again, if end users are not particularly sophisticated or experienced enough to find fault with this reasoning, they will accept the recommendation; and the resin selection will have been made by molders based on what is simpler for them.

Economics

Injection molders are not always motivated by an incentive to provide their customers with parts that have the ultimate in properties. When resin selection is granted to molders by their customers, they are likely to think in terms of providing parts that yield with the best profit to themselves, consistent with supplying adequate parts.

Costs involved in designing and making molds have already been alluded to. Obviously, a cut-and-try approach to mold making can be expensive; and if molders want to make the whole project look attractive to their customers, they do not want to frighten them off with excessive mold costs, nor do they want to defray these costs themselves.

Beyond the mold costs, other processing costs come into focus. Certainly molders are interested in optimum cycles in the operation of their machines. From their experience, they normally know which resins run well in their equipment so that they can achieve good cycles. This is not to imply that the average custom molder is not curious, or is disinterested in exploring new materials. There are times, however, when the custom molder's workload precludes the extra time required to become familiar with the processing characteristics of resins they have never run before. Frequently, they have had years of favorable experience with a limited selection of resins with good properties, and will want to provide the new part in one of these. Given implied authority by their customer, they will use the one that will result in the best cycles with the fewest problems, provided that it is a reasonable selection, although not necessarily optimum as far as the part's actual use is concerned.

A concomitant cost consideration is the cost of the resin. Unless the job the molders are bidding will call for truckload quantities of resin, they will be forced to lay in smaller (and therefore less economical) quantities. However, if they can suggest a resin they already use for other jobs, they can buy larger quantities and thereby save themselves a price/volume penalty in the cost of raw material. Then they can include the savings generated in this way as part of their profit. This is simply another example of how the molder's parochial considerations can effect resin selection.

8

Prototype Parts

INTRODUCTION

An early step in the conversion of a design idea into a production part is to obtain a limited number of prototype parts. There are a number of reasons for this. Subjectively, simply to see and handle the proposed new part in plastic is instructive. It is also important to make as many tests on prototypes as possible to determine if the resin selected for the part will perform as expected. Although prototyping can be expensive, failure to do so can at times be more costly in the long run if failure of the part becomes obvious only after production parts are the first ones to be evaluated.

In plastics, prototypes can be machined from rod or slab stock, or injection molded in an inexpensive single-cavity mold made of aluminum, mild steel, or epoxy steel. A third method, and in many ways the best, is through the use of preproduction tooling. A fourth technique is available if the part to be prototyped in plastic is already being produced in a die-cast metal. Somewhat overlooked in the past, this method of prototyping in die-casting tools offers some special advantages.

COMPARISON OF PROTOTYPING METHODS

Machining from Stock Shapes

If the end user has in-house prototyping facilities, of course, the work should normally be done there. Progress of the machining work can be followed closely, and obstacles that may be encountered as the work proceeds can possibly be overcome by small changes in design that can be suggested on the spot. However, perfectly satisfactory prototyping can also be carried out in a good independent machine shop having plastics experience. In either case the rod or slab required can be obtained from a stock shape distributor.

Some of the advantages of machining prototypes are:

1. Compared to the other routes, machining makes possible a short lead time.
2. Normally, machining is the least expensive route if only one or two parts are required.

Offsetting these advantages are a few limitations:

1. No data on optimum molding conditions are obtained.
2. Parts with truly thin walls should not really be attempted. Large thin-wall sections are difficult to machine accurately in plastics.
3. Only natural colors are normally available in stock shapes.
4. Some end-use tests performed could be misleading, since machining intricate parts with adequate fillets and without stress risers is difficult.
5. It is not economical to prepare very many parts for evaluation using this method.
6. It becomes expensive if there is a need to evaluate prototypes from several candidate resins.

Molding in Temporary Prototype Molds

Most of the time, molding in temporary prototype molds is one of the best approaches for obtaining enough useful prototype parts. The molds can be made in aluminum, steel, or epoxy steel, and the choice of these three can depend completely on the complexity of the part, the need for fast response, and costs. Some injection-molding shops specialize in making prototype tools and producing low-volume parts. These should be consulted before making a decision on the mold material.

Some of the advantages of obtaining prototype parts from temporary or prototype molds are:

1. An adequate number of parts for testing can easily be produced.
2. Valuable information can be obtained about the molding conditions required.
3. In nearly all cases, the prototypes can be used for realistic end-use testing, including part size, impact resistance, and environmental studies. Further, meaningful end-use specifications can be based on data obtained from these parts.
4. Unlike prototyping by machining, there are no limitations on wall thickness in the temporary-mold approach, other than those normally involved in any injection molded part.
5. Samples can be molded from several grades of a given polymer, or even from several candidate resins if the final resin selection is still in doubt.
6. Samples can be prepared in colors for final evaluation prior to color selection.

As could easily be predicted, there are several disadvantages to the temporary-mold route:

1. Costs for cutting the mold and molding the parts can be significant, particularly if the part is a complex one.
2. Improvements to part design become limited due to the expense of mold modifications, or they could become impossible, necessitating a completely new prototype mold.
3. Although temporary molds, which are almost always single cavity, are more simply made than production molds, lead time using this approach is frequently longer than is desirable.

Two different kinds of prototype molds can be considered, an inexpensive and simple one or a one-cavity preproduction mold. The simple prototype mold is typically made of kirksite, aluminum, mild steel, or brass. It generally lacks cooling, venting, slides, and ejector pins. The advantages of such a mold are quick delivery time and low initial cost. Unfortunately, it does not provide the opportunity to duplicate production conditions in terms of molding cycle, shrinkage, or warpage. Further, it provides little information with regard to gate selection, location of cooling lines, proper runner design, and functioning of the mold.

In summary, more information can be obtained from a prototype mold than from samples prepared by machining, but considerably less than can be deduced from a preproduction mold.

Preproduction Tooling

Preproduction tooling is the ultimate approach to prototyping. It allows the duplication of production parts precisely with respect to shrinkage, warpage, molding cycle, and properties. Because it provides precise cavity dimensions, it is useful in proper gate selection, accurate mold construction, and in providing information for possible part design modifications. In general, it greatly minimizes problems in building the final production mold and in predicting the performance of parts. The mold duplicates the production cavity environment exactly, and thus indicates the performance of parts molded from one or more candidate resins.

It is not advisable to take shortcuts in the design of preproduction molds. They should include any slides, cams, ejectors, cooling, and provisions for venting—ideally, every feature of the anticipated production mold. On the other hand, part layout in the mold cavity can be made simple, without ribs, bosses, or contouring, although thought must be given to what *might* be needed in this regard. Over design of the cavity interior will only limit future options.

Obviously, there is no point in making a plastic part more complicated than necessary. For example, if it is not certain that ribs are necessary, the preproduction mold cavity should be made without them. It is much easier to recut a mold cavity than to have to build up sections afterwards. On the other hand, it is important that the mold design not preclude these additional part features. Ejector pins, slides, cores, and cooling lines, and mold environment requirements should be designed with this in mind.

It is also important that the preproduction mold contains all the features that would be designed around one cavity of the final production tool. The best preproduction molds exactly duplicate the cavity environment that will be used in production while providing room for modification. Thus, it is useful to design, for example, a four-cavity layout first and simply section off one-quarter of it for the construction of the prototype. Then cooling lines can be designed for the one-cavity prototype mold to resemble as closely as possible their layout in one section of the final four-cavity mold. In addition, the mold can be gated exactly as the final part will be gated. Again, as with ribs, if it is not certain that the best gating location has been chosen, the prototype mold can be designed to allow alternative gating arrangements. It might prove better, for example, to gate into the side, or perhaps three gates might be needed, or perhaps a center gate is the most advantageous. Providing for this kind of flexibility is simple when the mold is first being designed.

What kind of information is obtainable from the preproduction mold? If it is designed to duplicate conditions in the final production tool, probably *all* the information that is needed to indicate whether or not a part will be successful can be obtained, including the ability to select the best engineering plastics for the application. On a short-term basis, preproduction tooling probably is the most expensive of the various routes to obtain prototype parts, and it requires more time; but in the long run it usually is the preferred route.

As stated, preproduction tooling *usually* is the preferred route, but careful continuity in the follow-up program is required. Occasionally, information is lost, or simply overlooked, as the plastic development program evolves from the preproduction mold to the final production stage. It sometimes seems as if there is a great vacuum into which there disappears all the useful data between the prototype shop and those responsible for final production tooling.

The information gap is often very obvious. It is not uncommon to let out a prototype mold job to one source and then later ask for quotes on the production tool from another source. If precise records are not maintained from the prototype tool and applied to the production tool, many advantages of the preproduction tool approach have been lost. It is even essential to resin selection to understand and interpret all preproduction prototype information correctly, or all the effort and information collected cannot prevent production problems.

A Case History

The following actual example illustrates what can go wrong as prototype information is brought to bear on the final production of a part.

A relatively thin automotive part used to perform a critical function was given rigidity through four intersecting ribs and a flange along its semicircular edge. It ran in the preproduction mold without problems. The end user immediately mounted the parts on cars and tested them at high temperatures. It was thought to be a marvelous application, and one that should be rushed into production, since it was 20% less expensive than a comparable metal part. It was to be installed on new-model cars, and there was to be no metal backup part. During the final months before the part went commercial, it was asked why the flange was needed. Nobody seemed to know for certain. Therefore it was decided that, since only the shell was critical to the part's function, the flange could be eliminated. At the same time, the ribs were also removed in order to reduce part weight.

At this point, a four-cavity production mold was commissioned. Some weeks later, the resin supplier received a call saying there were problems with the *resin*. The parts were warping. They were needed at the assembly line in 10 days, when production was to start. Upon arrival at the molder's shop, the supplier found boxes full of useless, warped parts.

All combinations of molding conditions were tried, but none eliminated the warpage. The production part was then compared with the prototype part, and it was noticed that the rim and flange had been removed, along with the ribs. In addition, the gating had been moved to the edge because it was believed this would allow for a more convenient cavity layout.

While pointing out the differences from the prototype that would affect part performance and processing, the material supplier decided the most expedient thing to do was to recommend that the part be gated more like the prototype mold, although the exact original gate location could not now be duplicated in the production tool. The next morning, the mold was in place. Two toolmakers had worked all night to accomplish this. The mold had originally hardened to 58 Rockwell C, and the toolmakers had gone through several carbide tools trying to make the changes. Parts were molded with the gating changes, but little improvement was noted.

It was obvious that the original rim and ribs would have to be restored. The mold was shipped back to the moldmaker, where it was reworked with only two of the original four ribs. The end user still wanted to minimize weight in the part. A few days later, the mold came back. The molded part failed again. It was only then that agreement was reached to restore all the ribs—an expensive proposition, since the steel had been hardened.

The production line, however, had already started. Every 90 seconds, a car was coming off the assembly line, and was parked in the holding area for lack of the plastic parts.

The mold was once again returned to the toolmaker, but only one more rib was added. When the mold was run this time, there was an improvement, but it was a small one. It was now 10 days past the original deadline. Once again, the mold went to the toolmaker, this time to have all the original ribs restored. When it came back to the molder this time, it worked, and the parts were acceptable, although part weight was up. The total cost: 20 days lost, three round trips for the mold to the toolmaker, days of delays in shipping cars, and an expensive retrofit assembly operation.

This experience emphasizes a couple of points: A preproduction mold will accurately predict results for the final mold and part requirements, provided all prototyping data are *recorded and used* when designing the final production mold. Although this case history may stray from the subject of resin selection directly, it touches on a vital part of the resin selection process—prototyping.

Molding in Die-Cast Molds

Engineering plastics are now being used in markets reserved for die-cast metals. When exploring these new applications, the need often arises to prepare prototypes of the metal part for testing. If properly planned, these prototypes in plastics from the die-cast mold can save design and project lead time and trim mold development costs. Successful molding of plastic prototype parts in die-cast molds requires a thorough understanding of the benefits and limitations of this useful method.

Some of the advantages of using die-cast molds for prototyping plastic parts are:

1. A large number of parts can be produced without the expense of a plastic test cavity. The cost is much less than machined parts if large numbers are needed.
2. Valuable time can be saved that would normally be spent waiting for the construction of a prototype mold. This will provide prototype parts quickly for preliminary decisions.
3. Parts obtained can be helpful in the study of postmolding operations, feasibility and cost studies, and evaluation of postmolding machining and assembly techniques.
4. A prototype part from a die-cast mold will help designers "see and feel" their design in a different material and plan better for the plastic mold.
5. Design problems can be identified.
6. The die-cast mold can easily be converted back to its original purpose.

Offsetting these advantages, there are several disadvantages:

1. Many metal die-cast parts are overdesigned, with unnecessarily thick wall sections, and frequently with poor fillet and radius design. The excessive thickness may lead to longer cycles than necessary, and lower impact toughness in the plastic prototype parts may result from the poor filleting.
2. Plastic shrinkage is usually higher than metal shrinkage, resulting in dimensional differences between the metal and the molded prototype plastic pats. This can be corrected in the final part design, however. Dimensional differences can also result from nonuniform mold cooling of the plastic and the effect of gate location. The anisotropic nature of some candidate resins may also result in some distortion or warpage of the molded part.
3. In-depth end-use testing of plastic parts produced in die-cast molds is usually not recommended, especially when impact or thermal cycles are to be evaluated. Erroneous test results and unnecessary engineering costs may result.

9

Estimating Part Costs

INTRODUCTION

Early in the resin selection process, the end user or designer will have reviewed the physical property requirements of a polymer for a proposed application, as well as its predicted performance in the environment to which it will be exposed. As mentioned, this normally results in finding at least several polymers that might be satisfactory. One of the next steps toward finalizing the selection is evaluating the estimated part cost for several candidates.

It is a rare end user that can estimate part costs accurately enough to coincide with the molder's estimated costs obtained later as the selection process proceeds. In fact, such accuracy is not really practical or necessary. On the other hand, the end user should be able to distill a multiple choice down to a few candidates based on estimated part cost, unless there is some unlikely situation where part cost is no object.

No doubt there are a number of possible procedures for preparing part cost estimates, but nearly all of them require the same input data. For this discussion, du Pont's cost estimator form[104] is a suggested method of determining the economic feasibility of a new design with different materials. It is simple and compact and, if used properly, will provide part cost estimates that can be confidently related to target part cost.

Keep in mind that this is a screening process only. If the part cost estimate obtained using the cost estimator form is within 20% of the target part cost, the next step is to consult an injection molder for a price quotation using the resin indicated by this procedure to be the most economical (always assuming that the selected resin has the required properties.)

EXAMPLE: COMPARISON OF ESTIMATED PART COSTS

This example should help you learn to use the cost estimator form, which is illustrated in Figure 9.1. For the example calculation, it is assumed that an end user's preliminary selection of a candidate resin to replace a metal part is based largely on tensile strength and stiffness as well as part thickness restrictions, and that the range of requirements of the part is covered by 30% glass-reinforced nylon 6/6, 30% glass-reinforced polycarbonate, and 20% glass-reinforced Noryl.®* Other properties of the three resins are assumed to be adequate for the proposed application, and part price becomes the leading consideration.

The base data used in example calculation are:

	30% GF[a] Nylon	30% GF[a] Polycarbonate	20% GF[a] Noryl
Material price per pound, $	1.70	2.03	1.89
Material specific gravity	1.30	1.43	1.27
Part weight, oz	3.76	4.13	3.68
Part wall thickness, mils	125	125	125
Number of parts	1MM	1MM	1MM

[a]GF means glass fiber.

The example (imaginary) part is presented as a flat plaque 5 in. × 8 in. × ⅛ in. thick. This geometry is selected for simplification, with the realization that in real life a part is seldom so simple.

*Noryl is a registered trademark of General Electric Co. It is a polyphenylene ether-based resin.

DESCRIPTION OF PART:

A RESIN PRICE ($/lb.)
B SPECIFIC GRAVITY
C PART WEIGHT
Equation for determining part weight in plastic, given the part weight in metal.
D WEIGHT PER 1,000 PARTS
E MATERIAL COST
F CYCLE
Request optimum cycle from appropriate resin supplier
G NUMBER OF CAVITIES
The equation supplied will provide an estimate of the minimum number of cavities that will be needed in one tool to meet your yearly part requirements.
H PARTS/HR.
I PROJECTED PART AREA
Area of the part when viewed, perpendicular to the plane of the parting line.
J TOTAL PROJECTED PART AREA
This is the area taken up by all of the shot when viewed as above.
K CLAMP NEEDED
The force, in tons, that the molding machine clamp must exert to overcome the force separating the mold halves caused by the pressure of the melted polymer during injection.
L SHOT SIZE
Weight of resin required to fill the cavities and runner system.
M MACHINE SIZE
This equation selects a machine that will likely provide adequate melting capacity assuming reciprocating screw equipment. For ram machines substitute 1 for ¾ in the equation.
N MACHINE CLAMP CAPACITY
O MACHINE HOUR RATE
P PROCESSING COST
The cost of processing is adjusted to reflect:
95% Processing Yield—some of the parts must be reground and remolded because they are out of tolerance or are damaged during molding.
80% Machine Utility—a portion of the machine down time must be assigned to each part molded.

	DU PONT ENGINEERING PLASTICS		
A Resin Price ($/lb.)			
B Specific Gravity			
C Part Weight (ounces) Plastic Part Weight $= \frac{\text{Sp. Gr. (plastic)}}{\text{Sp. Gr. (metal)}} \times$ Metal Part Weight			
D Part Weight (lb./1000 units) $\frac{(C)}{16} \times 1{,}000$			
E Material Cost($/1000 units) $\frac{(D) \times (A)}{0.95}$			
F Cycle (sec.) from Figure 1 or Figure 3			
G Number of Cavities No. Cav. = (No. of parts required annually) $\times$ (F) $\times$ 10^{-7} (see note 1) Round off to 1 or nearest higher even digit.			
H Parts/Hour $\frac{(G) \times 3{,}600}{(F)}$			
I Projected Part Area (in.^{2})			
J Total Projected Part Area (in.^{2}) (I) $\times$ (G)			
K Clamp Needed (tons) 5 $\times$ (J) (Note 2)			
L Shot Size (oz.) (C) $\times$ (G) $\times$ (W) The shot size factor (W) from Figure 9.2 is an estimate of material in the sprue and runners.			
M Machine Size (oz.) 0.75 $\times$ (L) $\times$ $\frac{60}{(F)}$ (note 3)			
N Machine Clamp Capacity (tons) Compare machine clamp vs. machine size (M) in Figure 9.3 with clamp needed (K) and select the larger of the two.			
O Machine Hour Rate ($/hr.) Select machine hour rate from Figure 9.4 corresponding to the machine clamp capacity selected in (N). (Note 4.)			
P Processing Cost ($/K) $\frac{(O) \times 1{,}000}{(H)\,(0.95)\,(0.80)}$			
Q TOTAL MANUFACTURING COST ($/K) = (E) + (P)			

FIGURE 9.1 Cost estimator form. (From Ref. 104.)

(A) *Resin price*. In this example, the resin prices were an average of the market prices listed in the March 1996 issue of the *Plastics Technology Resin Price Update*.[105]

(B) *Specific gravity*. Obtained from product literature.

(C) *Part weight* (oz). For this example, it was assumed that an aluminum part was to be replaced. The specific gravity used for the aluminum was 2.77, and the original part weight in aluminum was 8 oz.

(D) *Part weight (in pounds per thousand units).* As shown in Figure 9.1.

(E) *Material cost (in dollars per thousand units).*

(F) *Cycle (molding cycle in seconds).* For approximate cycles to use in this example of three different resins, the appropriate material suppliers were contacted and estimates obtained.

(G) *Number of cavities.* Determined from calculation as shown in Figure 9.1.

(H) *Parts per hour.* Determined from calculation as shown in Figure 9.1.

(I) *Projected part area.* The example part was 5 in. × 8 in. in area.

(J) *Total projected part area.* Determined from calculation as shown in Figure 9.1.

(K) *Clamp needed (tons).* Determined from calculation as shown in Figure 9.1.

(L) *Shot size (oz).* Determined from form calculation. *W* is read from Figure 9.2, and includes sprues and runners.

(M) *Machine size (oz).* Determined from calculation as shown in Figure 9.1.

(N) *Machine clamp capacity (tons).* In this example, the clamp need from K (at 800 tons) exceeded that obtained from M in Figure 9.3, so the larger value was used.

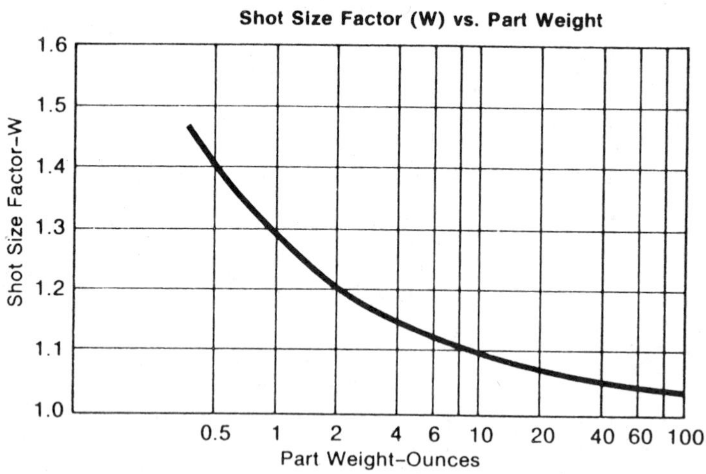

FIGURE 9.2 Graph showing shot size factor *W* versus part weight. (From Ref. 104).

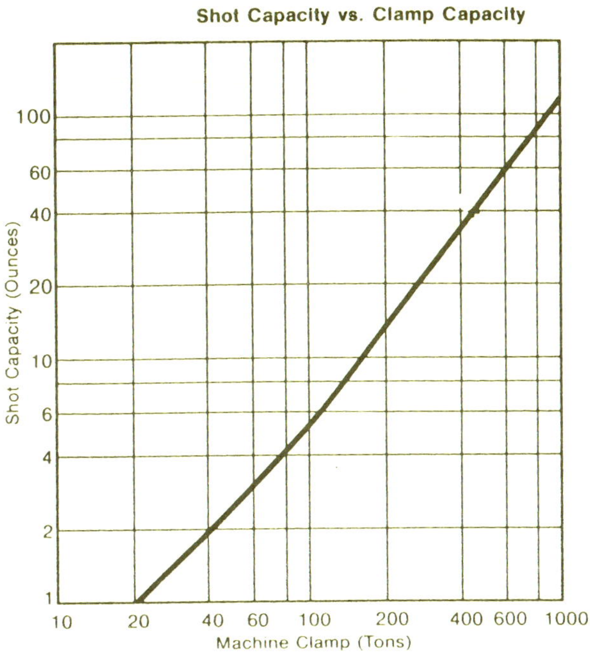

Figure 9.3 Graph of shot capacity versus clamp capacity. (From Ref. 104.)

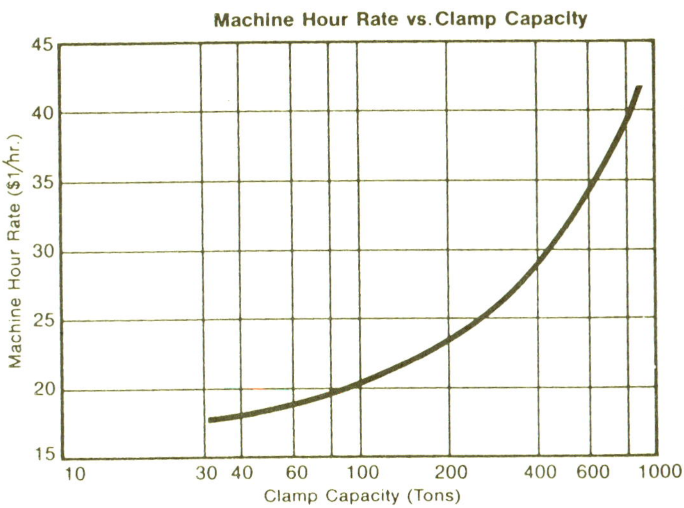

Figure 9.4 Graph of machine hour rate versus clamp capacity. (From Ref. 104.)

(O) *Machine hour rate ($/hr).* Obtained from an 800-ton clamp machine from Figure 9.4.

(P) *Processing cost ($/hr).* Determined from calculation as shown in Figure 9.1.

(Q) *Total manufacturing cost ($/1000 parts).* Determined from calculation as shown in Figure 9.1.

Of the three resins used in this example calculation, the glass-reinforced nylon 6/6 has the lowest cost, some 30% below the cost for the glass-reinforced polycarbonate. This difference should be considered significant, and emphasis placed on using a resin with such a price advantage. However, the data being entered into the form must be as accurate as possible to avoid distortion. For example, if instead of estimating the cycle for polycarbonate to be 25 s, a 20-s cycle had been used, the cost difference between the nylon and polycarbonate would have been reduced to about 26%.

SUMMARY

One of the benefits of the early cost estimating exercise on the part of the end user, in addition to a comparison of part costs for different raw materials, is to get some concept of the economic feasibility of making a profit from the proposed plastic part. If estimated costs are within the end user's objective, they should then obtain more precise costs from one or more injection molders.

Another benefit of the cost estimating effort is to eliminate any gross cost disadvantage that might evolve from a resin under consideration. As an example, it is possible from the cost estimation to discover that a candidate resin based on price and specific gravity might be attractive, only to find that a long molding cycle makes part costs prohibitive.

In summary, the cost estimation exercise using the form is simple and requires little time. It can bring out significant cost differences and problems, but can also be misleading unless the data used are reasonable.

10

Review of Key Engineering Thermoplastics

COMPOSITIONS SELECTED FOR INDIVIDUAL ATTENTION

As has been stated frequently throughout this book, it is impractical to list everyone of the hundreds of available engineering plastics compositions, yet the purpose of the book is to come to grips with some kind of viable selection process. As the properties of the polymers are reviewed, it can be seen that it is misleading to state that one of the major engineering plastics is always to be preferred to the others in a particular environmental or use situation, or that one resin usually selected for its particular usefulness in a single type of application is limited to that application.

Realizing these limitations, the approach here is to restrict the number of resins discussed in detail to a relative few (15), with the idea that they are representative enough of the whole spectrum to illustrate the thought process that goes into resin selection. The selected group has been chosen largely on the basis of the relative volumes used in the molding industry, although some exceptions have been included made to illustrate certain unique properties one polymer may contribute to the flexibility of the resin spectrum.

The 15 resins selected cover a broad spectrum of physical properties of engineering thermoplastic resins, and form a pool from which a very large percentage of applications of engineering plastics can be accommodated.

Another consideration is that most of the resins have a number of glass-reinforced, UV-stabilized, heat-stabilized, mineral-reinforced, flame-retarded, etc., versions; and it would be unwieldy to include all of them. However, reinforcement of some of the polymers with glass or minerals greatly enhances their versatility; and for this reason, some of the reinforced versions are individually included in the ranking in Chapter 11. Specifically, some of the 30% (approximate) glass-reinforced versions of nylon 6/6, polycarbonate, and polyester are covered individually because of their importance in the family of engineering plastics; while glass-reinforced compositions of acetals, polypropylene, and ABS are not included because the bulk of these reins are sold in unreinforced form. Mineral-reinforced compositions of nylon and polypropylene are included in the ranking because of their widespread use, particularly in automotive and housing applications.

Within the group discussed in detail, it must be understood that all have high value-in-use properties for many applications; and frequently three or four of them will be technically adequate for a given use. At the same time, each polymer does have at least one Achilles heel; and there are occasions when only one of the resins is really adequate for a given application.

In this chapter, then, brief descriptions of the resins are provided, including chemical structure, a discussion of properties, discussion and illustrations of some typical end-use applications, and special considerations when they are important.

ABS

Chemistry

ABS is actually a terpolymer obtained from the polymerization of acrylonitrile and styrene monomers in the presence of butadiene rubber (Figure 10.1).

Because ABS is a three-component system, many variations are available, some for general purposes, some for various levels of higher impact resistance, others for the ease of plating. It therefore is difficult to enumerate the pros and cons of ABS in anything other than general terms. For the purpose of this dis-

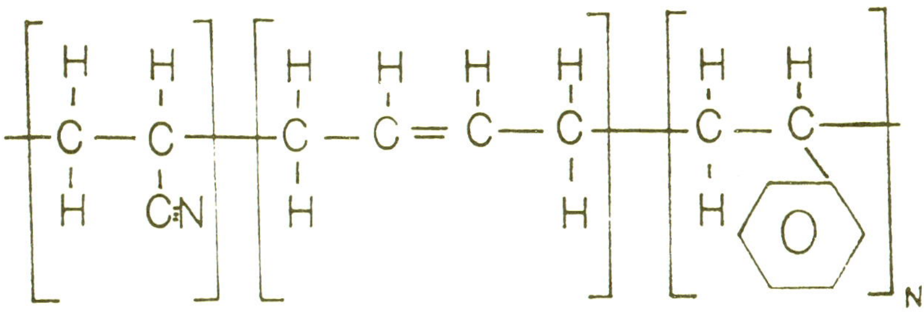

FIGURE 10.1 Chemical structure of ABS terpolymer.

[illegible], a medium-impact resin has been considered, and physical property data are reported that only approximate typical properties.

Properties

ABS is a versatile material with a number of good properties, and is, with the exception of polypropylene, the least expensive of the resins included in this book's limited list of engineering plastics. In a sense, however, ABS barely makes it into the loose definition assigned here for engineering plastics, because of its limitations at elevated temperatures, where its physical properties fall off sharply. Useful properties of ABS include:

Good stiffness
Excellent toughness
Good electricals
High gloss
Good chemical resistance
Platability

End uses

Pipe and fittings are the largest market for ABS, particularly in drain, waste, and vent pipes. In terms of what is thought of here as an engineering plastic,

ABS is also widely used for telephone housings, business machine housings, appliance housings, and automobile grills (either painted or electroplated).

Because of its high gloss and good appearance, as well as its moderate cost, it is widely used for luggage, sports accessories, and many other items, where end-use temperatures are below 200°F.

Typical application

An important use of ABS as an engineering plastic is in automotive instrument panels. Figure 10.2 is a photograph of a Ford Thunderbird panel, where the ABS was used because of its low cost relative to other materials that were adequate for the use, its impact resistance, and its ease of painting and decorating.

Acetals

Chemistry

There are two large-volume commercial version of acetal resins, one a homopolymer of formaldehyde, the other a copolymer of trioxane, the cyclic trimer of formaldehyde (Figure 10.3).

Each of these two acetals is manufactured by a different major domestic resin supplier, and there are some measurable differences in the physical and

FIGURE 10.2 Ford Thunderbird instrument panel in ABS. (Courtesy of Borg-Warner Chemicals, Inc., Parkersburg, WV.)

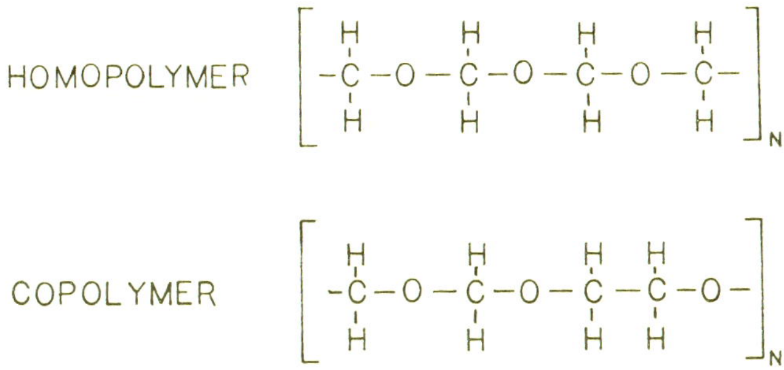

FIGURE 10.3 Chemical structure of acetal homopolymer and copolymer.

mechanical properties between the two. Separate property summaries are given later in this book. However, for the purpose of discussing resin selection, the term "acetal" is meant to include them both.

Properties

Acetals are versatile, with a number of useful properties:

- Good mechanical strength and rigidity
- Excellent fatigue resistance
- Abrasion resistant
- Low coefficient of friction
- Resistant to moisture, gasoline, solvents
- Resistant to repeated impacts
- FDA approval for many food-contact applications
- Low creep

End uses

Acetal resins find uses in many industries. The automotive industry, for example, use acetals for carburetor and fuel tank applications, partly because of its resistance to gasolines and oils.

It finds applications in the appliance and machinery industries because of its fatigue resistance and low coefficient of friction. The plumbing industries, particularly sprinkler systems, use acetal widely because of its dimensional stability in water. Acetals are also used in environments where food contact, stain resistance, high gloss, and resistance to extended hot-water immersion are involved.

Special considerations

Acetals have normal shrinkages in the range of 20 mils/in., higher than for many other engineering plastics. Thus, to design a complex mold to allow for these shrinkages and obtain a part with the required dimensions requires a certain amount of expertise.

Although acetal resins have excellent resistance to many organic and inorganic materials, they are attacked by strong acids such as hydrochloric and sulfuric acids.

Typical applications

Figure 10.4 is a photograph of a ballcock assembly injection molded from an acetal homopolymer. Over 200 million of these assemblies are in use, performing well over the years. Acetal was selected for this application because of its dimensional stability in water, its fatigue resistance, stiffness, strength, and low-wear characteristics.

Figure 10.5 is another example of a typical end use for acetals. The large gears for the appliance were put in acetal because of its excellent fatigue resistance, high tensile strength, low-wear property, and dimensional stability.

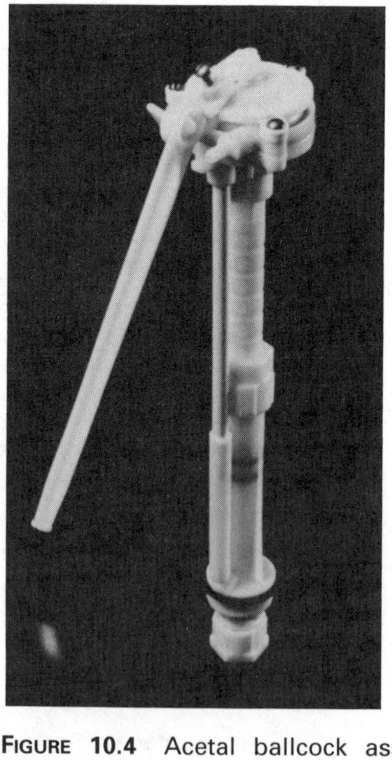

FIGURE 10.4 Acetal ballcock assembly. (Courtesy of E. I. du Pont de Nemours & Co., Wilmington, DE.)

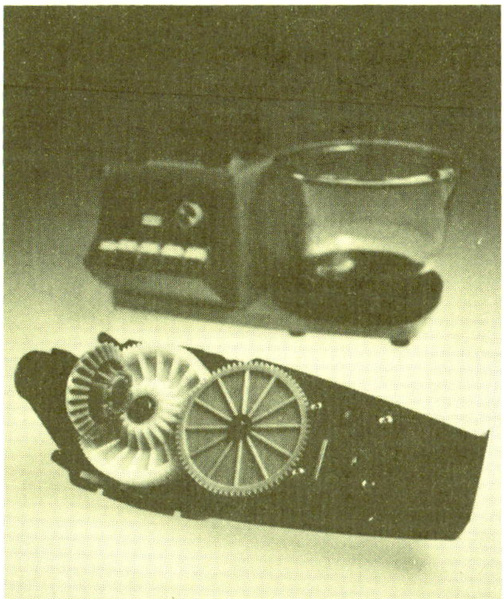

FIGURE 10.5 Acetal gear assembly. (Courtesy of E. I. du Pont de Nemours & Co., Wilmington, DE.)

Nylon 6

Chemistry

Nylon 6 type is a homopolymer, a condensation polymer of ∈-amino caprolactam (Figure 10.6).

Properties

The properties of nylon 6 are similar to those of nylon 6/6, but not exactly the same (see, for example, Tables 1-5–1-7). (The important physical property data for both nylons are shown in a succeeding chapter.)

Toughness at relatively high and low temperatures
Resistance to repeated impacts
Resistance to abrasion and wear
Resistance to attack by many organic solvents, oils, and gasolines

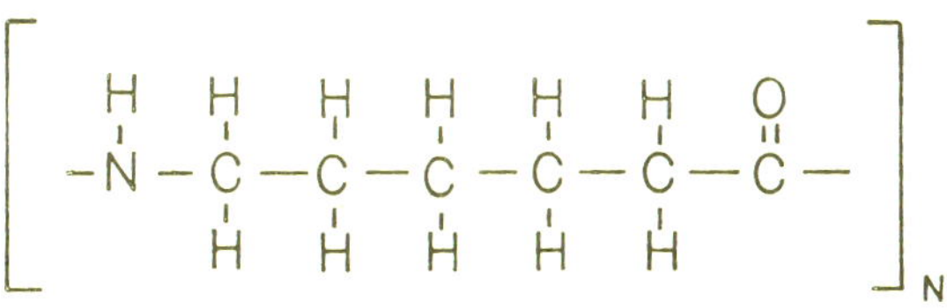

FIGURE 10.6 Chemical structure of nylon 6 polymer.

Whereas nylon 6/6 moisture conditioned at 50% relative humidity and absorbs 2.5% water, nylon 6 under the same conditions absorbs 2.8%. While these differences are real, if proper designing is employed, this dimensional growth can usually be accommodated.

At the 2.8% moisture level, nylon 6 has greater impact resistance than nylon 6/6, and greater elongation. On the other hand, nylon 6/6 has a higher modulus, lower creep, and a higher melting point.

Nylon 6/6

Chemistry

Nylon 6/6 is a condensation polymer of adipic acid and hexamethylenediamine (Figure 10-7).

Properties

Nylon is an older, established engineering plastic, long noted for its wide use throughout a number of major industries. The reason for its versatility is its combination of properties, including:

Toughness at high and low temperatures
Resistance to repeated impacts
Retention of stiffness, tensile properties, and shape at high temperatures
Resistance to abrasion and wear
Resistance to attack by many organic solvents, oils, and gasolines

End uses

Nylon is used extensively in the automotive industry in a number of critical areas, such as in the engine, transmission, cooling system, and emission controls, where toughness and long life are required at temperatures of 200–300°F or above. With toughness, abrasion resistance, and repeated impact resistance inherent in the polymer, nylon is used extensively for cams and gears in hardware, plumbing, appliance, and machinery applications. It is also used for many purposes in the communication and electrical industries.

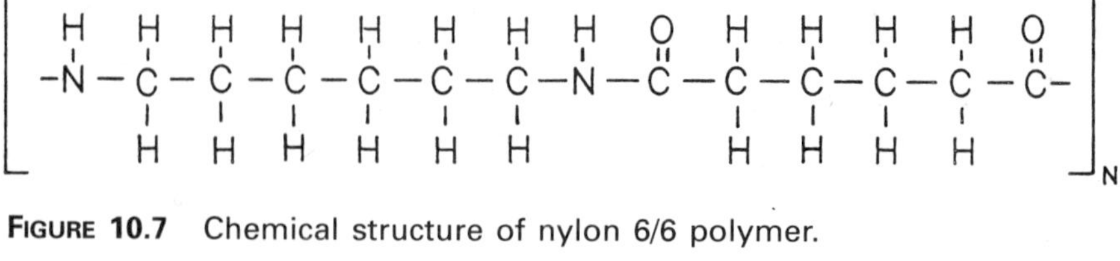

FIGURE 10.7 Chemical structure of nylon 6/6 polymer.

Another reason for its almost universal use is that, over the years, a broad family of hundreds of versions has evolved that is adapted, usually through the use of additives, to fit many specific conditions. Examples of this include 6/6 resins with special antioxidants, UV stabilizers, heat stabilizers, hydrolytic stabilizers, etc. If plastics designers feel that most of the requirements needed for a proposed application point in the direction of nylon, they should consult material suppliers to be certain the version they are selecting is the best one.

Special considerations

The principal consideration to be made when evaluating the use of nylon for an application is the effect that water absorption may have on it. The nylon 6/6 family (as well as nylon 6) will gain about 2.5% by weight when conditioned to equilibrium moisture content at 50% relative humidity. In addition to dimensional considerations, other properties are affected as nylon picks up moisture from the time it is molded (usually containing about 0.2% water at that point in time) until it has reached moisture equilibrium. For example, yield stress drops off about 29%, elongation increases about five times, toughness about doubles, while stiffness decreases about 57%. All of these factors must be carefully considered in the design of nylon parts. Over the years, so many thousands of successful applications for nylon have accumulated that a significant amount of experience and design data have become available that are useful to the designers, and materials suppliers are eager to supply them.

Nylon offers excellent resistance to a wide range of organic and inorganic substances, but as is the case for any polymer, there are agents that will attack it, particularly under conditions of stress. Considerable caution should be used, and preliminary tests made, when the proposed environment is likely to include exposure to strong acids or certain salt solutions.

Typical applications

Figure 10.8 shows several take-off, driven, and governor gears molded from a nylon 6/6. The abrasion and wear-resistant properties, as well as significant cost and weight reductions (from the metals previously used) were the motivating forces in this application. These are some of the early engineering applications, dating back to 1954.

Figure 10.9 illustrates the use of nylon 6/6 in an electromechanical function. Nylon has many uses of this kind because of its combination of strength and stiffness, toughness, chemical resistance, heat resistance, and resistance to abrasion. (Nylon is usually restricted to power applications of less than 500 V at normal power frequencies.)

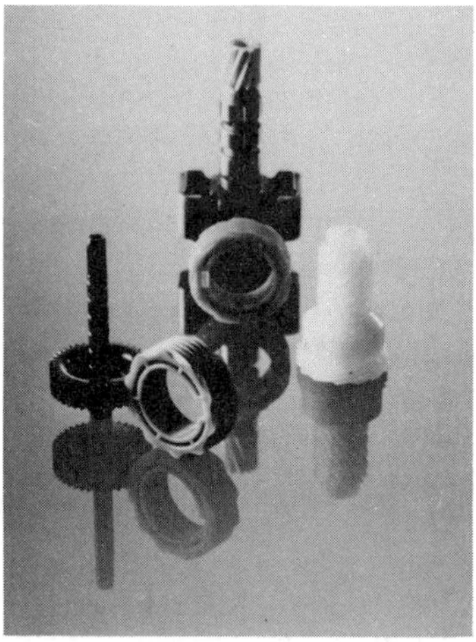

FIGURE 10.8 Nylon gears. (Courtesy of E. I. du Pont de Nemours & Co., Wilmington, DE.)

Glass-Reinforced Nylons

Description

As might be expected, nylons can be obtained with quite a few different glass compositions. Possibly the largest volume glass-reinforced nylon contains 30–33% glass, and that is the one included in these comparisons.

Properties

Major property improvements of glass-reinforced nylons over the unreinforced versions are:

Greater tensile strength
Greater stiffness
Better impact strength
Greatly improved creep resistance
Better dimensional stability
Enhanced resistance to sustained high temperatures

End uses

Because it remains stiff even at 250°F, glass-reinforced nylon 6/6 has a growing number of automobile under-the-hood uses, such as valve covers, fan blades, and power steering reservoirs.

Glass-reinforced nylon 6/6 provides excellent fatigue resistance, and is used successfully as springs and gears at high-stress levels. Because of its en-

FIGURE 10.9 Nylon coil forms. (Courtesy of Celanese Plastics and Specialties Co., Chatham, NJ.)

hanced dimensional stability and impact resistance, it is finding its way into housings for power tools and into mechanical applications such as plumbing spuds and nuts, and mechanical couplers.

Special considerations

Like the parent polymer, glass-reinforced nylon 6/6 absorbs moisture, but measurably less, and with less direct effect on properties. The weather resistance of glass-reinforced nylon is superior to the base polymer, although it is still not recommended for long-term outdoor exposure without adequate UV stabilizers.

In terms of chemical or environmental effects, the glass-reinforced nylons remain essentially the same as a base resin.

Warpage obtained with these resins is somewhat greater than with the base resin, particularly in thin section, but the shrinkage is less.

Typical applications

Figure 10.10 is a photograph of thrust washers and oil filter tubes molded from a glass-reinforced nylon 6/6 composition. These parts are used in automobile

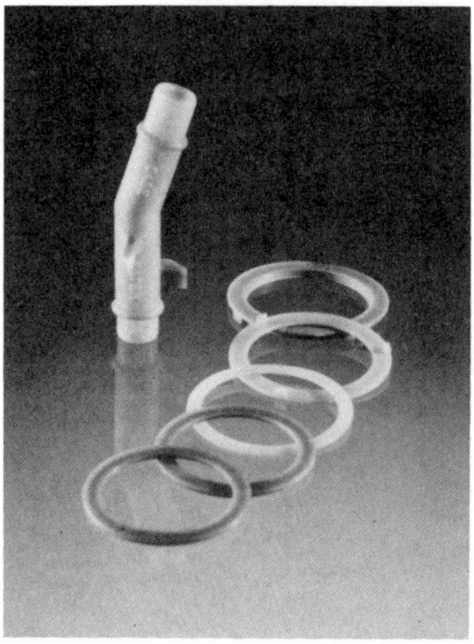

FIGURE 10.10 Glass-reinforced nylon thrust washers and an oil filter tube for automobiles. (Courtesy of E. I. du Pont de Nemours & Co., Wilmington, DE.)

transmissions, and have been in operation for over 10 years, indicating their resistance to attack by hydraulic fluids. Low friction, wear resistance, and outstanding creep resistance to loads up to 300 lb at 2000 rpm were some of the reasons for selecting the nylon for the thrust washers.

Figure 10.11 shows a key housing for an automobile, molded from a glass-reinforced nylon 6/6. The cost of the part was significantly less than that of the die-cast part it supplanted, and it weighs a quarter of a pound less. The glass-reinforced nylon provides the high stiffness and strength required for this exacting application.

Mineral-Reinforced Nylons

Description

There are many varieties of mineral-reinforced nylon 6/6 and nylon 6 on the market, from a number of suppliers. They differ in the kind of mineral used and the percent of the mineral. Some have glass fiber reinforcement as well, some have toughening agents, stabilizers, etc., each designed for fairly specific end-use properties. Again, a problem arises as to which one to include in the comparisons and ranking, and a 40% mineral-filled nylon 6/6 has been arbitrarily selected for that purpose.

The 40% mineral-reinforced nylon 6/6 composition is indeed reinforced by

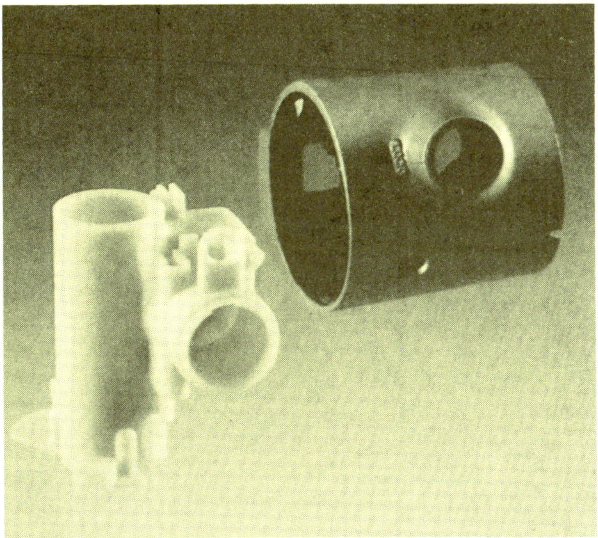

FIGURE 10.11 Glass-reinforced nylon automobile key housing. (Courtesy of E. I. du Pont de Nemours & Co., Wilmington, DE.)

the addition of the mineral, not just filled; that is, there is a chemical bond between the nylon and the mineral particles, thus enhancing the tensiles and stiffness over those of the base resin. Although the example considered here is stiffer than the base resin, it is not as stiff as the 30–33% glass-reinforced compounds.

Properties

Some of the important property improvements of mineral-reinforced nylons over unreinforced nylons are:

Greater tensile strength (but less than glass-reinforced nylon)
Greater stiffness
Better dimensional stability
Improved molding cycle

End uses

Basically, mineral-reinforced nylons have been developed to provide a resin with the intrinsic good properties of nylon polymer, but with enhanced stiffness, dimensional stability, low warpage, and property retention at elevated temperatures. Glass-reinforced nylons provide higher stiffness, tensiles, etc., as has already been noted; but they warp more than the mineral-filled resins, and they are more expensive.

Because warpage is usually acceptable and the surface of the molded parts is very good, mineral-reinforced nylons find many uses where a fairly large surface area is involved, such as some of the smaller automobile body panels. It is also used in engine parts such as carburetors, fuel pumps, and fan shrouds, and in such items as handles for hardware items, computer spools, and appliance parts.

Special considerations

As is the case with the glass-reinforced resins, mineral-reinforced nylon absorbs less moisture than the base polymer, with less direct effect on properties. It has essentially the same resistance (or lack of it) to chemical or other environmental conditions as the base resin.

Typical applications

Figure 10.12 illustrates the use of a mineral-filled nylon 6/6 in carburetors. The material was selected because of its dimensional stability and warp resistance, as well as its resistance to gasoline and gasohol.

Figure 10.13 is a photograph of a climate-control support bracket for an automobile, molded in a mineral-filled nylon 6/6. Formerly in metal, the switch

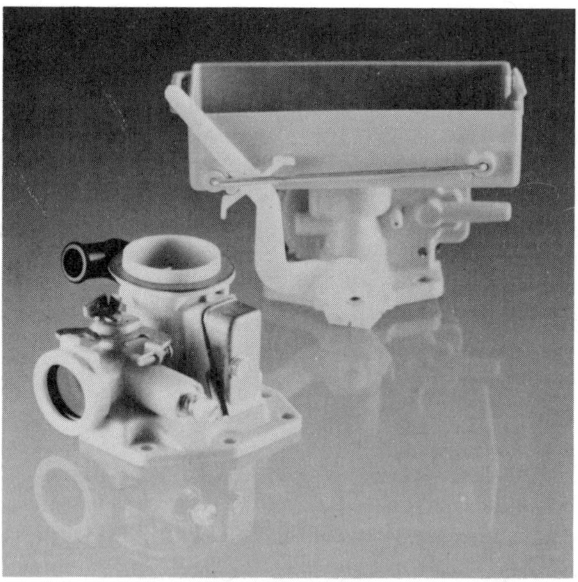

FIGURE 10.12 Carburetor of mineral-filled nylon. (Courtesy of E. I. du Pont de Nemours & Co., Wilmington, DE.)

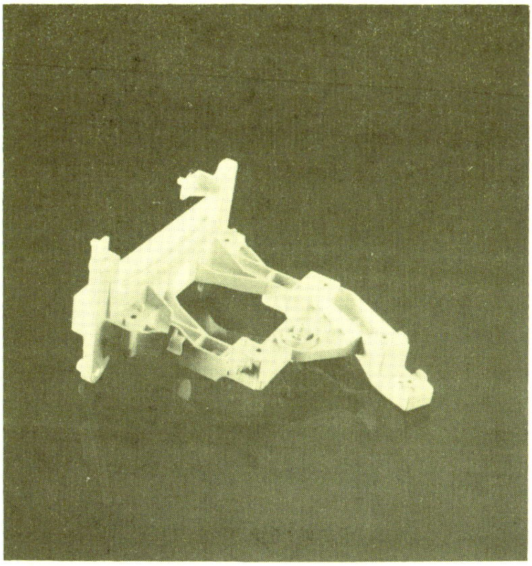

FIGURE 10.13 Support bracket of mineral-filled nylon. (Courtesy of E. I. du Pont de Nemours & Co., Wilmington, DE.)

was made to plastic to reduce weight. The mineral-reinforced resin was selected to provide warp-free, dimensionally stable, stiff, strong parts.

Toughened Nylon (Super Tough®)

Description

Toughened nylon compositions usually contain an elastomericlike toughening agent in a matrix of base polymer. Fairly new to the family of engineering plastics, their unusual toughness has broadened the horizon for plastics and made them unique resins worthy of separate consideration.

Several suppliers provide toughened nylons, but there are differences between the resins that make it impractical to deal with them generally. For this reason, Du Pont's Zytel Super Tough nylon 6/6 is used as a model in this discussion.

Properties

The outstanding toughness of Zytel Super Tough® is, of course, its most noteworthy property. In addition to toughness, it has:

Fast molding cycles
Good surface and colorability

Good tensiles
Exceptional resistance to crack initiation and propagation

Unusual for polymers, the toughened nylon has exceptional resistance to both crack initiation and crack propagation, even with low-moisture content and regardless of notch radius.

End uses

One of the first to take advantage of the impact resistance of toughened nylon was the sports equipment industry. Some sports equipment benefiting from the high-impact properties of the toughened nylon include roller skates, ice skates, racquets, bicycle wheels (glass reinforced), and ski boots and bindings. Industrial power tool housings are a natural for the tough resin, as are industrial fertilizer spreaders and other farm machinery.

Automotive uses for the toughened resins include radiator fans, trailer hitch covers, and body parts under the car subject to flying gravel, etc.

Special considerations

The toughened nylons do not have the same degree of stiffness or creep resistance as the base 6/6 polymer, nor quite the tensiles. They are subject to essentially the same chemical and environmental considerations.

Because of the similarity of toughened 6 and 6/6, there are many applications where either can be used. However, the stiffness, tensiles, and yield strength for nylon 6 are inferior to these properties in 6/6; so there are areas where 6 loses the competitive race to 6/6. For example, there are some under-the-hood applications where 6/6 is preferred because of its higher heat distortion properties; while in some kinds of gears, bearings, couplings, etc., type 6 may be used or even preferred.

Because of its tendency to absorb more moisture than nylon 6/6, nylon 6 requires even more care during part design to allow for the dimensional changes from the as-molded part, as well as the somewhat lower stiffness and yield stress.

Since 6 and 6/6 nylons are so similar chemically, nylon 6 is subject to the same special consideration as 6/6; that is, exposure to certain agents as some salt solutions and strong acids can lead to problems, and preliminary testing in such environments is necessary.

Typical application

Figure 10.14 shows a 30-in.-wide hood extension for a row-crop tractor molded from a toughened nylon 6/6. The toughened nylon was required to resist breakage in normal incidents involved in farm work. It is also flexible enough to

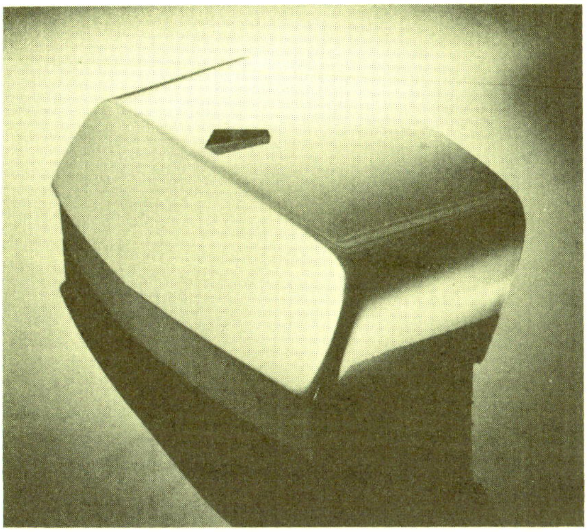

FIGURE 10.14 Hood extension for tractor in Super Tough nylon. (Courtesy of E. I. du Pont de Nemours & Co., Wilmington, DE.)

mate smoothly with an irregular hood, the creep resistance to remain bolted in such a position, and resistance to solvents and chemicals used in farming.

Polycarbonate

Chemistry

Polycarbonate thermoplastic resin is a condensation polymer of phosgene and bis-phenol (Figure 10.15).

Properties

Polycarbonate resins span a wide range of physical properties when versions containing glass-reinforcing fibers or flame retardants are included. Perhaps

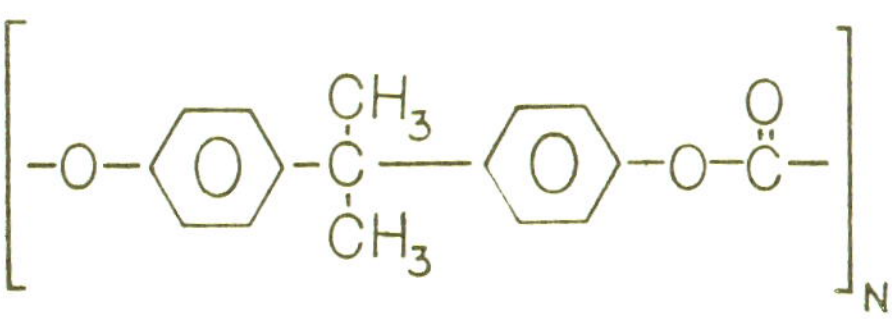

FIGURE 10.15 Chemical structure of polycarbonate.

polycarbonate's most recognized property, particularly to the general public, is one of toughness, as illustrated by fastball pitchers trying to blow a pitch through a sheet of the material. And it is true that polycarbonate is tough. In addition, polycarbonate has other useful properties:

Dimensional stability
Low mold shrinkage
Transparency
Low creep
Good electricals
Low warpage

End uses

Because of its low shrinkage, low warpage, and dimensional stability, polycarbonate is often used for intermediate-sized housings, such as automotive bezels, appliance housings, and electronic structural components. Polycarbonate has found its way extensively into the electrical/electronic industries, and is also used in structural components in electrical applications. Because it is an amorphous material, it is used in molding applications where the combination of transparency and strength is an important criterion for an application.

Special considerations

Polycarbonate is a very useful polymer, but a number of considerations need to be taken into account before its selection as the material for an application is finalized.

Toughness of polycarbonate can be a function of thickness and temperature. When a wall thickness of ¼ in. is exceeded, some of the unique impact properties of polycarbonate are diminished; and when temperatures are below 0°F at any thickness, the impact values are sharply reduced. Even in these conditions, however, polycarbonate remains as tough as some of the other engineering plastics. Available product data must be consulted.

While polycarbonate resin generally is stable to water, mineral acids, and organic acids, crazing and/or embrittlement may occur if a part molded from the resin is exposed to hot water or a moist, high-temperature environment. As a result, a temperature limit of 140–160°F is recommended under these conditions. Further, certain other chemical environments can also adversely affect polycarbonate; and, for this reason, lubricants, gaskets, o-rings, cleaning solvents, etc., must be evaluated.

Molded-in inserts are not recommended for use with polycarbonate, particularly in applications exposed to thermal cycling, although it may perform adequately with glass-reinforced versions because of the lower coefficient of thermal expansion and higher stress limit. Press-fits are possible with polycarbonate, but very high-stress levels leave the part more susceptible to chemical and thermal attack. Ultrasonic methods are probably the best routes for inserts for polycarbonate.

In an application where fatigue resistance is a significant property, special attention to design in polycarbonate is necessary, and prototype testing should be thoroughly undertaken.

Typical applications

Figure 10.16 illustrates a common use of polycarbonate as an electrical tool housing. Low shrinkage permits molding to fine tolerances, and the polymer is dimensionally stable. In this particular example, impact strength is also an important consideration.

Figure 10.17 illustrates the specific useful property combination of transparency and toughness.

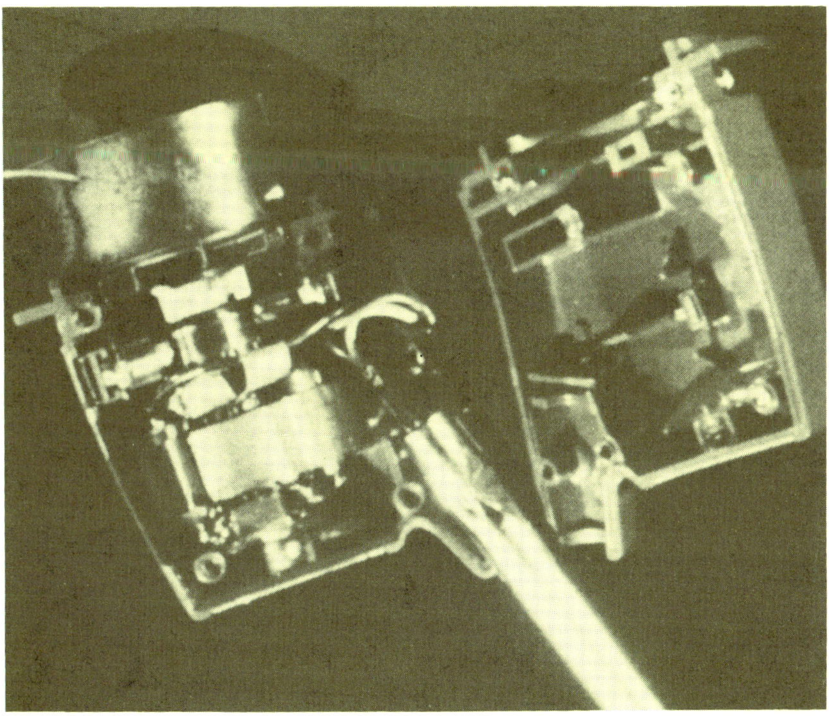

FIGURE 10.16 Electrical tool housing in polycarbonate. (Courtesy of General Electric Co., Pittsfield, MA.)

FIGURE 10.17 Transparent vessel of polycarbonate for home appliance. (Courtesy of General Electric Co., Pittsfield, MA.)

Glass-Reinforced Polycarbonate

Description

Glass reinforcement in polycarbonate results in the expected improvements in mechanical properties over the base polymer, along with some offsetting reduction (In this discussion, only a 30% glass level is considered). Some of the differences are:

Higher tensile strength
Greater stiffness
Better creep resistance
Reduced impact resistance
Better fatigue endurance
Improved dimensional control through low shrinkage
Loss of transparency

End uses

Glass-reinforced polycarbonate is used largely in the same basic applications as the base polymer, such as electrical/electronic components and appliance housings, but where higher tensiles or stiffness are needed.

Special considerations

Those used to the unusual toughness of the base polycarbonate resin should be cautioned to consider the reduction in impact resistance caused by the addition of glass. At the same time, a dramatic improvement in fatigue endurance occurs as a result of the glass addition,which can be an important property improvement.

The glass-reinforced versions of polycarbonate are, in general, subject to the same special considerations as the base resin, although molded-in inserts are possible with the glass-reinforced material, in contrast to the cautions spelled out for the base resin.

Polyphenylene Ether-Based Resin (Noryl®)*

Chemistry

In the family of engineering plastics, one of the major members is a polyphenylene ether-based resin made by General Electric with the trademark name of Noryl. A component used to modify the processing characteristics and properties of the base resin is proprietary. The chemistry of the polyphenylene ether, the major component, is shown in Figure 10.18.

Properties

Noryl has a broad spectrum of uses, including many in the electrical/electronics industries. Some of its key properties are:

Good impact resistance
Low mold shrinkage
Good dimensional stability
Good electricals
Low warpage

End uses

Because of its low shrinkage, dimensional stability, and low warpage, PPE-based resin has found many uses in the production of housings for appliances

* Noryl® is a trademark of General Electric Co. It is also referred to as PPE-based resin in this book.

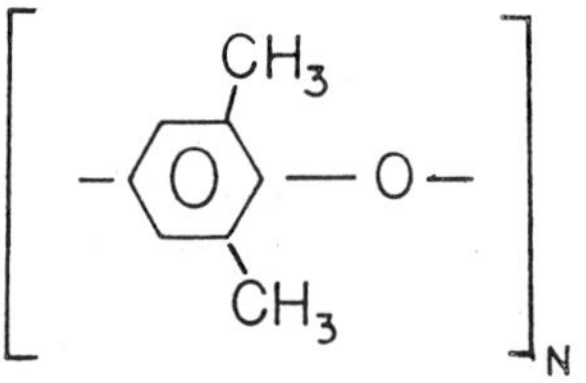

Figure 10.18 Chemical structure of polyphenylene ether, base resin for Noryl.

and business machines. It is used in the electrical industry in switches, cable connectors, and wiring device components. In the electronics fields, it is used for such television componentry as yokes, yoke supports, connectors, and coils.

Special considerations

PPE-based resin has excellent hydrolytic stability, and is more resistant to acids and bases than some of the other engineering plastics. However, it is very susceptible to attack by halogenated and aromatic hydrocarbons. Other organics, such as ketones and esters, also affect the resin, and careful consideration should be given to proposed uses in which the parts are to be exposed to any organic solvent.

Typical application

Figure 10.19 shows a number of typical items molded from PPE-based resin. Business machine and computer housings are molded from this resin for a number of reasons, including excellent processibility into large area parts, good surface, low shrinkage, and excellent dimensional stability.

PPE-Based Resin—Glass Reinforced

As is the case with all plastics, glass reinforcement of PPE-based resin improves some important properties of the base resin at the sacrifice of a few others. For example, the glass-reinforced versions offer a 35°F increase in the heat-distortion temperature of the base polymer, plus a threefold increase in tensiles and flexural modulus. Mold shrinkage decreases at least 50% from the base polymer, which is desirable, but the notched Izod impact decreases 50%, which is not, although the glass-reinforced material still has a reasonably high degree of toughness.

Properties

To summarize the major changes ingoing from the base polymer to a 30% glass-reinforced version, the glass-reinforced resin has:

FIGURE 10.19 Housings molded from PPE-based resin. (Courtesy of General Electric Co., Pittsfield, MA.)

Higher tensile strength
Greater stiffness
Improved creep resistance
Reduced impact resistance
Better dimensional control through lower shrinkage

End uses

The glass-reinforced modified PPE-based resins are typically found in electrical connectors, structural parts, television components, liquid-handling equipment, and business machine applications.

Special considerations

Like the base polymer, the glass-reinforced versions have excellent hydrolytic stability, along with greater strength and stiffness. At the same time, the glass-reinforced material is also susceptible to attack by halogenated and aromatic compounds. In addition, ketones and esters can affect the properties to some extent, and care must be exercised if the proposed environment includes any organic solvent.

Polypropylene (Talc Filled)

Chemistry

The polypropylene resin referred to in these discussions is a 40% talc-filled homopolymer. The simple molecular structure of the polymer is shown in Figure 10.20.

Properties

Of the many polypropylene compositions available in the marketplace today, only the talc-filled version is covered here because several properties of the unfilled polymer, particularly its heat-deflection temperature, do not meet the arbitrary standards set up here for engineering plastics; and its other properties are not unique enough to overcome this deficiency. However, when polypropylene is filled properly with talc to a percentage of 40%, its physical properties are enhanced, and permit some engineering plastic applications not subject to heat-deflection limitations.

A typical (40%) talc-filled polypropylene has the following characteristics:

Good stiffness
Fair toughness
Moderate tensile strength
Excellent resistance to chemical and solvent attack
Good surface on molded parts
Low cost

End uses

Talc-filled polypropylene has a variety of uses in a number of industries, but typical uses include automotive fan shrouds, automotive air conditioning conduits, some appliance housings, and some electrical applications.

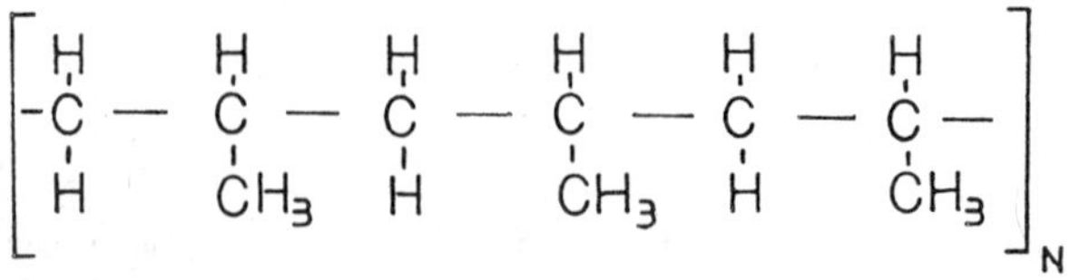

FIGURE 10.20 Chemical structure of polypropylene.

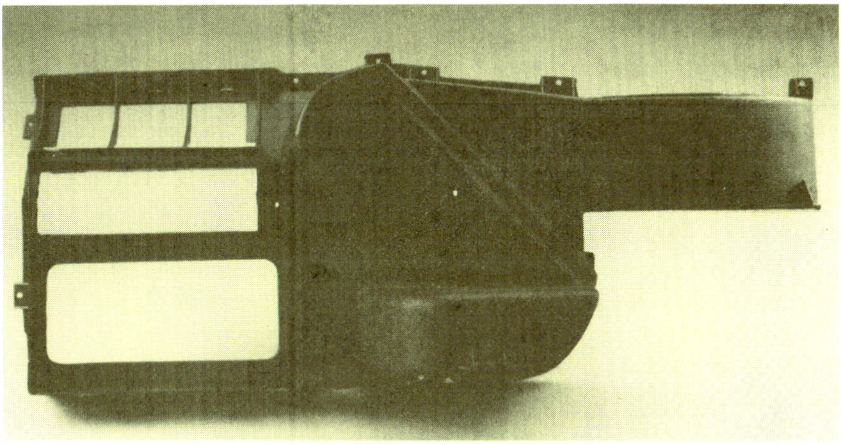

FIGURE 10.21 Automobile heater housing in talc-filled polypropylene. (Courtesy of Hercules, Inc., Wilmington, DE.)

Typical application

Figure 10.21 shows an automobile heater housing in a talc-filled polypropylene. The polypropylene composition is lightweight, stiff, adequately heat resistant, and less expensive than most other candidates for this application.

Polysulfone

Chemistry

Polysulfone is a condensation polymer of dichlorodiphenyl sulfone and bisphenol (Figure 10.22).

Properties

Polysulfone is a transparent engineering plastic with a number of good properties. Some of the important ones are:

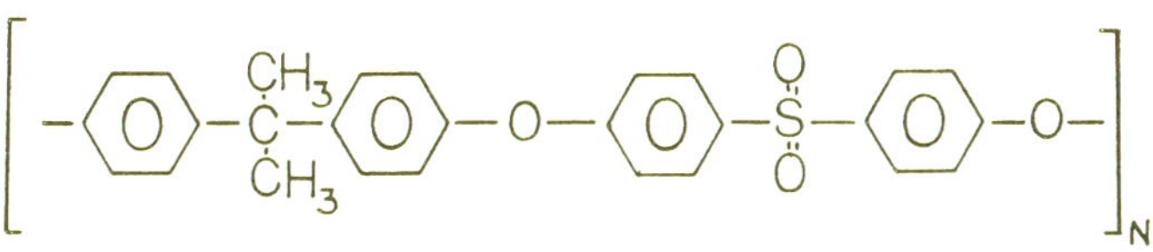

FIGURE 10.22 Chemical structure of polysulfone.

Retention of properties up to 300°F for long periods of time
Excellent resistant to effects of steam
Good creep
Good electricals
Resistant to radiation
Resistant to acids and bases

End uses

Because of its unique resistance to effects of steam, polysulfone has found a number of applications in the medical industry for surgical tool housings, humidifiers, respirators, and a number of other laboratory equipment items. With FDA approval in hand, polysulfone is also used in food processing equipment and sanitary pipe, milking machines, and filtration membranes and housings.

Because it is more expensive than some of the other engineering plastics, it is used for the more exotic electrical applications such as certain under-

FIGURE 10.23 High-frequency circuit board in polysulfone. (Courtesy of Union Carbide Corp., Danbury, CT.)

the-hood components in automobiles, printed circuit boards, terminal blocks, and stereo components. Its unique property of photochemical inertness lead to its use in a number of parts of "instant" cameras.

Special considerations

Although polysulfone is highly resistant to mineral acids, alkali, and salt solutions, and reasonably resistant to detergents and hydrocarbon oils, it is readily attacked by polar solvents such as ketones, chlorinated hydrocarbons, and aromatic hydrocarbons. Those responsible for resin selection should take these latter effects into consideration, since they represent a fairly wide range of possible contaminants in a variety of environments.

Typical applications

Figure 10.23 presents an example of a high-frequency circuit board molded from polysulfone. Polysulfone was selected for its excellent high-temperature electrical properties, and it is being used frequently to replace the more expensive cooper-clad laminates.

Figure 10.24 is a typical illustration of an end use for polysulfone. The vessel is a suction bottle used in hospitals for the removal of unwanted body

FIGURE 10.24 Typical laboratory vessel in polysulfone. (Courtesy of Union Carbide Corp., Danbury, CT.)

fluids. The bottle has the necessary property of transparency. It is tough, and can stand the vacuum applied to it in use, and will remain free of crazing and cloudiness when exposed to steam sterilization procedures.

Thermoplastic Polyesters

Chemistry

As is the case with acetals, there are two polymers in the engineering plastics category that are both polyesters, having similar properties. They are the condensation polymers of dimethyl terephthalate with 1,4-butanediol or with ethylene glycol. The polybutylene terephthalate (PBT) is currently the larger volume of the two, and its chemical structure is shown in Figure 10.25.

Although there are commercial versions of PBT and PET in the unreinforced, mineral, mineral-glass and glass-filled states, it is the 30% glass product that is the most widely used and, hence, it has been selected for specific treatment here.

The chemical structure of polyethylene terephthalate (PET) is shown in Figure 10.25.

The 30% glass-reinforced version of PET has properties similar to the PBT, although stiffer and with higher tensiles; however, in general, the PBT and PET compete for the same end uses, and they are discussed here as a single family.

POLYBUTYLENE TEREPHTHALATE (PBT)

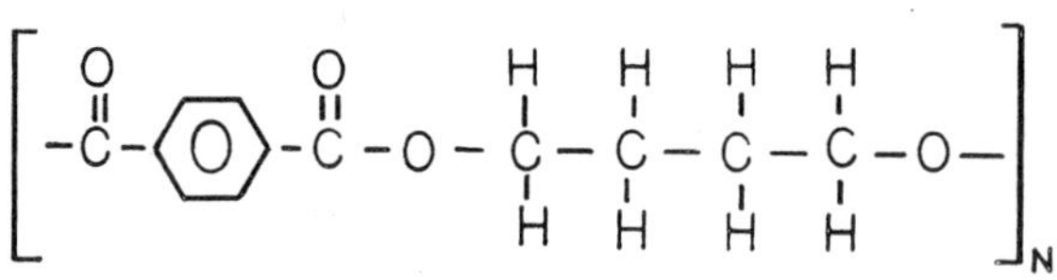

POLYETHYLENE TEREPHTHALATE (PET)

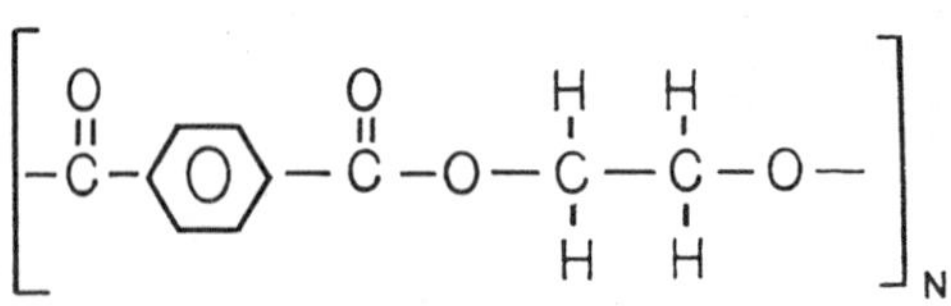

FIGURE 10.25 Chemical structure of polybutylene terephthalate and polyethylene terephthalate polyester polymers.

Properties

Polyesters are newcomers to the plastics engineering scene, but they have been fast growing due to the combination of a reasonable price and excellent properties. The desirable properties of the polyesters are:

Excellent dimensional stability
High strength
High stiffness
Low creep at elevated temperatures
Excellent electrical properties
Good chemical resistance
Low mold shrinkage

End uses

Because of the combination of good electrical and heat-resistance properties, polyesters are used under the hood in the automotive industry for ignition system components such as distributor caps, coil bobbins and rotors, and light sockets. Other automotive uses include painted exterior body panels and louvers, transmission system components, and emission-control system parts.

The use of polyesters in the electrical/electronics industry is large and growing, with such applications as electronic connectors, switches, relay cases, and covers. They are also used in the appliance industry for handles, small housings, impellers, and general engineering parts.

Special considerations

Although the polyesters have stability in the presence of water at normal ambient temperatures, prolonged exposure at temperatures higher than 125°F are not recommended. Prolonged exposure to acids and bases, particularly at elevated temperatures, will result in polymer deterioration.

Some reinforced polyester compositions demonstrate significant anisotropic behavior, and warping of moderately long and/or thin parts is something that must be considered in the early stages of resin selection and part design.

As is the case with some other plastic-families (e.g., carbonates, arylates), polyester resins must be very dry during the molding process. The best results will be obtained when the moisture content of the resin in the hopper is <0.02%; and to accomplish this, a high-efficiency hopper dryer is essential.

Typical applications

Figure 10.26 illustrates a common use for polybutylene terephthalate (PBT) polyester. Outstanding long-term creep resistance at elevated temperatures is a

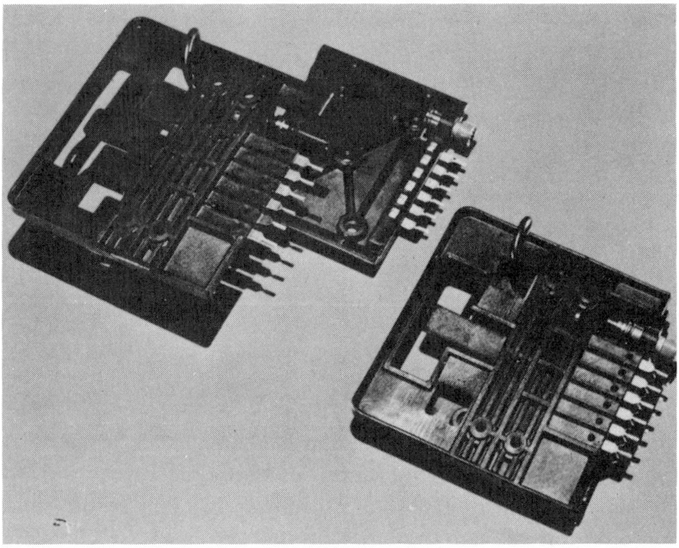

FIGURE 10.26 Automobile radio bracket in PBT polyester. (Courtesy of General Electric Co., Pittsfield, MA.)

FIGURE 10.27 Part of high-energy ignition system in PBT polyester. (Courtesy of General Electric Co., Pittsfield, MA.)

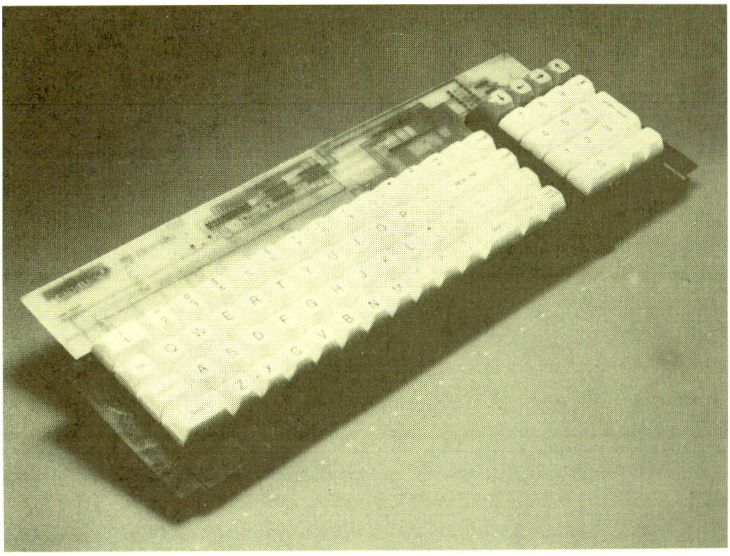

FIGURE 10.28 Computer keyboard in PET polyester. (Courtesy of E. I. du Pont de Nemours & Co., Wilmington, DE.)

major requirement of a material to be used in this quick-connect radio mounting bracket.

Figure 10.27 contains an example of a part used in an automotive high-energy ignition system. There is a trend toward replacement of the thermosets formerly used in this application with thermoplastic polyesters. The polyesters are tougher, and offer high resistance to arc tracking in high-voltage use. Their high-heat-deflection thermal stability make the polyesters an excellent choice for under-the-hood electrical applications.

Figure 10.28 is a good example of an application for polyethylene terephthalate (PET) polyester. It was used in this computer keyboard primarily because of its outstanding stiffness and dimensional stability as well as its good electrical insulating properties.

11

Ranking of Resins and Selection Process

INTRODUCTION

The ranking of polymers is based on a reasonable selection of the more important physical properties, those that are discriminating and considered important by designers and others involved in the resin selection process.

In the special case of the various nylon compositions, the values used, when available in the literature, are those reported on specimens conditioned to 50% relative humidity and 70°F, since this is thought to be more realistic than using dry-as-molded values, which can change significantly due to moisture pickup. As mentioned previously, all material suppliers do not seem to obtain the same physical property data for compositions that are supposed to be identical, and sometimes subtle differences among versions of material suppliers are reported. As a result, when doubts arose in the preparation of the ranking data, some averaging of the values was done, so that the values reported may not be the same as those officially presented in the literature of material suppliers. When the resin selection progresses to the point of looking in depth at just a few compositions, the decision maker is at liberty to consult the product literature more thoroughly for more accurate values to include in design calculations.

In any event, when more than one supplier manufactures a resin that appears to be a reasonable candidate, it is definitely advisable to contact at least two of them. While all material suppliers are very responsible organizations, in the world of subtle differences in products (when that is the case in deciding between two or more candidate resins), it is human nature to be somewhat subjective in describing the merits of a proprietary resin.

If none of the resins in the ranking appears to be suitable, and most of the material suppliers contacted do not have a truly reasonable candidate, it is possible that one of the newer, more "exotic" polymers that are less well-known will be precisely what is needed. Most of these possibilities are discussed briefly in Chapter 14.

Anyone who is at all familiar with properties of engineering plastics can appreciate the degree of difficulty of even attempting to rank a number of them according to their properties in relation to each other. It is, in fact, impossible to do so with precision. As a simple example, when two polymers have identical values for a property, obviously, one cannot be ranked better than the other based on that property alone. As a more involved example, one resin may have a lower cost per cubic inch than another but may require a longer molding cycle, thus nullifying the apparent advantage. A number of other anomalies crop up when ranking is attempted.

Considering all these caveats, it is hoped that the reader will accept the rankings on the basis on which they are presented; that is, they are simply a rough guide to aid resin decision makers and can serve only to eliminate some obviously inadequate candidates and point out a group of several candidates deserving more consideration.

To proceed, in each case, the ranking order within each physical property category is headed by the composition having the "best" value for that particular property, and the remainder of the compositions follow in descending order.

As an example, it would normally be very unusual to select a resin based on one property alone, such as stiffness; but if that situation arose, the stiffness ranking would indicate that, of the 15 resins, the glass-reinforced PET is the stiffest, followed by PBT, etc. If an application requires both a high degree of stiffness and toughness, review of the rankings on toughness might suggest the use of glass-reinforced polycarbonate, since it ranks high in both stiffness and toughness. On the other hand, if the application calls for stiffness, toughness, *and* resistance to an active ketone, ranking of resins to families of chemicals and solvents would suggest that a glass-reinforced nylon might be preferred, since it is high in stiffness, has reasonable toughness, and is not affected by ketones.

Using such a process for the various physical properties demanded by a proposed application, usually at least two or three of the resins will appear as possible choices, and preliminary design calculations can then be made to confirm their suitability. It is these resins that should be followed up more thoroughly by discussions with the appropriate material suppliers for more in-depth data and with molders for their input on processibility, etc. For example, estimated molding cycles for the required part thickness can then be obtained, and of course, the prices involved. Part cost estimates can then be made for the candidate compositions (see Chapter 9), and the resin selection process can proceed to the next step.

In the event that none of the specific resin compositions provided in the rankings appears to be attractive as presented, the selector will want to review with material suppliers their line of products and modifications of the ranked resins, for more often than not, a compromise can be made this way.

For example, a resin choice from these ranked resins will not be a VO composition, and if that is needed, the suppliers must be contacted to determine if a VO version is available (it almost always is). Or the application may demand a weatherable composition with special resistance to a continuous high-heat environment, and a special UV- and heat-stabilized version is available for these requirements. Many such modifications exist, which accounts for the hundreds of resin variations available in the marketplace.

It is unfortunate that sufficient useful data common to all the ranked resins for such important physical properties as fatigue endurance, resistance to repeated impact, abrasion resistance, and frictional properties are not readily available. Such data are not reported at all for some of the resins, and those that are reported at times are done so on different bases that cannot be translated to a common denominator.

In any case, it is possible that none of the ranked resins, or their modifications, will be acceptable for a particular use. In that event, many choices have already been removed by reference to the ranking, but the search must continue, and will involve researching for the more exotic and specific polymers.

Even a cursory review of the summary of physical properties of the ranked resins (see Table 11.1) will show what appears to be a marked superiority in structural properties of the glass-reinforced resins over the base resins. However, they are more anisotropic than the unmodified resins. Depending on the geometry of the part, values of some physical properties can therefore be

TABLE 11.1 Summary of Physical Property Values of Ranked Resins

Resin	Tensile strength (kpsi) ASTM D-638	Flexural modulus (kpsi) ASTM D-790	Heat-deflection temperature (°F @ 264 psi) ASTM D-648	Notched Izod (ft. · lb/in.) ASTM D-256	Elongation at break (%) ASTM D-638
ABS	5.0	370	216	6.0	5–25
Acetal					
Homopolymer	10.0	400	277	1.4	40
Copolymer	8.8	375	230	1.3	60
Nylon 6[a]	8.8	150	170 (DAM)[b]	2.6	300
Nylon 6/6[a]	11.2	180	194 (DAM)[b]	2.3	300
Nylon 6/6[a] (30% glass)	20.0	900	485 (DAM)[b]	2.0	5
Nylon 6/6[a] (mineral reinforced)	9.0	600	446 (DAM)[b]	1.0	10
Nylon 6/6[a] (Super Tough)	7.5	130	159 (DAM)[b]	40.0	215
Polycarbonate	9.0	340	270	15.0	110
Polycarbonate (30% glass)	19.0	1100	295	2.0	4
PPE-based resin	9.6	360	265	5.0	60
PPE-based resin (30% glass)	17.0	1100	300	2.3	5
Polypropylene (40% talc)	4.6	490	170	0.4	8
Polysulfone	10.3	400	345	1.4	75
Polybutylene terephthalate (30% glass)	17.3	1100	405	1.8	3
Polyethylene terephthalate (30% glass)	22.3	1300	435	1.9	3

Creep modulus (kpsi)	Dielectric strength (V/mil)[c] ASTM D-149	Arc resistance (s.) ASTM D-495	Specific gravity ASTM D-792	Approximate melting point (°F)	Approximate average shrinkage (mil/in.) ASTM D-955	Cost (¢/in.^{3})[d]
215	420	83	1.05	230	8	3.4
250	500	220	1.42	397	22	6.3
250	500	240	1.41	329	22	6.3
63	400 (DAM)	65 (est.) (DAM)	1.13	420 (DAM)	15–20	5.9
83	420 (DAM)	116 (DAM)	1.13	491 (DAM)	15–20	6.5
600	420 (DAM)	96 (DAM)	1.30	500 (DAM)	6–10	9.8
380	430 (DAM)	139 (DAM)	1.51	493 (DAM)	9	6.3
39	380 (DAM)	131 (DAM)	1.10	491 (DAM)	15–20	7.7
320	380	10	1.20	302	5–7	6.7
730	475	10	1.43	302	2–3	10.4
335	550	75	1.06	265	5–7	5.2
1300	550	9	1.27	265	1–3	10.7
450	430	142	1.22	333	10	2.3
350	425	120	1.24	392	7	20.9
750	475	146	1.53	437	4–8	10.0
1000	550	135	1.56	489	4–6	7.4

[a]At 50% RH, unless otherwise specified.
[b]DAM is dry as molded.
[c]⅛-in. thickness.
[d]1996.
Source: Ref. 105.

significantly less in the transverse flow direction than in the direction of flow. In complex parts where weld line formation during molding cannot be avoided, significant decreases in tensile strength, for example, can be expected, occasionally down to values reported for the unreinforced material itself.

This is not to say that reinforced resins are not reliable, for current experience indicates that the volume of sales of such products is increasing rapidly. However, a high proportion of new applications do not demand the very highest stiffness possible, nor the strongest composition, nor the best in creep that are offered by the glass-reinforced versions. Instead, the largest volumes consumed are generally in the middle of the ranked property ranges, where the unreinforced resins can provide perfectly adequate design properties without the concern about machine wear and lack of esthetic appearance that the reinforced resins frequently entail. By adding ribs or thickening crucial wall sections, strength and stiffness of the base resins can be enhanced enough to provide satisfactory parts without the need for reinforcing additives. In summary, reinforced resins have a number of properties superior to those of the base resins, but in general, there should be some compelling end-use or design reason for selecting them in preference to the base resins.

It is interesting to realize that a good designer can very often use any one of several candidate resins, provided it is compatible with the proposed environment. Nevertheless, it is hoped that those referring to this book for assistance in resin selection will be able to save time arriving at reasonable candidates, and will be able to proceed with the final decision on resin selection with confidence.

Finally, it should be noted that it is ultimately end users who usually must assume the responsibility for the final selection of an engineering plastic. Injection molders can help considerably in arriving at a final selection, and material suppliers can help by actually recommending a specific resin. In the final analysis, however, end users must decide which material to use and normally must be prepared to assume responsibility for its success or failure after they have made appropriate end-use tests.

RANKING BY RESISTANCE TO CHEMICALS AND SOLVENTS

Probably the very first consideration in the resin selection process is *environment*, including effects of exposure of the proposed part to unusual temperatures, weather, excessive moisture, and solvent and chemicals. With a given resin, little if anything can be done through part design to offset the effects of chemicals or solvents.

There are an almost unlimited number of chemicals and solvents that might be involved in an end-use application, and obviously it is not reasonable to have test data on the effect of all of them on the engineering plastics. In the *Modern Plastics Encyclopedia*,[106] the response of most of the resins ranked here to a limited number of common chemicals and solvents is included. Those not included have been covered from other sources.

As a preliminary screening reference, Table 11.2 summarizes the resistance of the ranked resins to inorganic acids and bases, to polar salt solutions, and to organic solvents, including aromatics, gasolines, alcohols, ketones, and aldehydes.

For a more accurate reflection of the facts of environmental resistance, either the material suppliers must provide the facts from their tests on their products, or the end users must test specimens in the proposed media when the information is not available from the suppliers. In any case, as the resin selection process proceeds to a final choice of candidates, they should be tested in the media. If the ranking data indicate attack by a solvent or chemical on one of the polymers being considered, it is wise to exclude it at the outset and to continue the search considering only those resins shown in the data that are not subject to attack by the chemical or solvent that is expected to be in contact with the proposed part.

RANKING BY TENSILE STRENGTH

Tensile strength is another property of polymers that can be reproduced accurately enough to permit realistic comparisons among resins, provided test conditions are carefully controlled. The ranking of the polymers in Table 11.3, then, can be used with some confidence. However, the resins do not behave identically as temperatures or load vary, and this must be kept in mind further along in the resin selection process.

For the most engineering plastics applications, tensiles below 5000 psi are usually inadequate, but ABS and polypropylene have been included in the list of 15 because of other properties that make these resins useful in a certain limited range of mechanical applications.

RANKING BY FLEXURAL MODULUS (STIFFNESS)

Stiffness is a property that is measured by a reasonably discriminating test that can provide a comparison between resins. It is often an important consideration in the selection of an appropriate resin; and ranking in terms of stiffness is a realistic exercise, keeping in mind that there are circumstances where design thickness may be increased to compensate for some differences in modulus.

Table 11.4 lists the engineering plastics chosen for the ranking process, with approximate modulus values obtained from various literature sources.

The data supplied for the nylons are reported for samples conditioned to

TABLE 11.2 Resistance to Ranked Resins to Chemicals and Solvents[a]

Resin	Strong acid	Strong base	Salt solution	Aromatics	Gas	Alcohol	Ketone	Aldehyde
ABS	N	N	N	A	SA	SA	SA	SA
Acetal	A	A	N	N	N	N	N	N
Nylon 6	A	N	SA	N	N	N	N	N
Nylon 6/6	A	N	SA	N	N	N	N	N
Nylon 6/6 (Super Tough)	A	N	SA	N	N	N	N	N
Nylon 6/6 (30% glass)	A	N	SA	N	N	N	N	N
Nylon 6/6 (mineral reinforced)	A	N	SA	N	N	N	N	N
Polyethylene terephthalate (30% glass)	N	SA	SA	SA	N	N	A	A
Polycarbonate	N	A	N	A	N	N	A	A
Polycarbonate (30% glass)	N	A	N	A	N	N	A	A
Polybutylene terephthalate (30% glass)	N	SA	SA	SA	N	N	N	N
PPE-based resin	N	N	N	A	A	A	A	A
PPE-based resin (30% glass)	N	N	N	A	A	A	A	A
Polysulfone	N	N	N	A	A	A	A	A
Polypropylene (40% talc filled)	N	N	N	A	SA	A	N	N

[a]In general, the indicated resistance to chemical or solvent attack is at room temperature. N, not affected; SA, slightly affected; A, affected.

TABLE 11.3 Descending Order of Tensile Strength Values for Ranked Resins

Tensile strength (kpsi) ASTM D-638	Resin	Rank
22.3	Polyethylene terephthalate (30% glass)	1
20.0	Nylon 6/6 (50% RH) (30% glass)	2
19.0	Polycarbonate (30% glass)	3
17.3	Polybutylene terephthalate (30% glass)	4
17.0	PPE-based resin (30% glass)	5
11.2	Nylon 6/6 (50% RH)	6
10.3	Polysulfone	7
10.0	Acetal	8
9.6	PPE-based resin	9
9.0	Nylon 6/6 (50% RH) (mineral reinforced)	10
9.0	Polycarbonate	11
8.8	Nylon 6 (50% RH)	12
7.5	Nylon 6/6 (50% RH) (Super Tough)	13
5.0	ABS	14
4.6	Polypropylene (40% talc filled)	15

TABLE 11.4 Descending Order of Modulus (Stiffness) for Ranked Resins

Modulus (kpsi) ASTM D-790	Resin	Rank
1300	Polyethylene terephthalate (30% glass)	1
1100	Polybutylene terephthalate (30% glass)	2
1100	PPE-based resin (30% glass)	3
1100	Polycarbonate (30% glass)	4
900	Nylon 6/6 (50% RH) (30% glass)	5
600	Nylon 6/6 (50% RH) (mineral reinforced)	6
490	Polypropylene (40% talc filled)	7
400	Acetal	8
400	Polysulfone	9
370	ABS	10
360	PPE-based resin	11
340	Polycarbonate	12
180	Nylon 6/6 (50% RH)	13
150	Nylon 6 (50% RH)	14
130	Nylon 6/6 (50% RH) (Super Tough)	15

50% RH and 73°F, a "normal" environment. This is a realistic assumption for many final end uses.

The actual as-molded stiffness for nylon 6/6, for example, is 410 kpsi, compared with 180 at 50% RH. Not many end uses lend themselves to constant low-humidity conditions that would retain the as-molded high-stiffness value, so it is up to designers and end users to determine which stiffness value they should design for.

RANKING BY FLEXURAL STRENGTH

Flexural strength is a measure of the amount of stress a material will withstand before it breaks under load as applied in the ASTM D-790 test. The reported flexural strength values are not often used in design calculations per se, and will not be used in the selection process, but can be useful to the selector in ranking qualitatively the 15 resins for their ability to resist stress levels before breaking. The ranking of those values (see Table 11.5) very closely approaches the ranking under flexural modulus (see Table 11.4).

Data for Super Tough are not available. Assumption is made here that Super Tough (at 50% RH) is so flexible and tough the sample will not break under test conditions of ASTM D-790.

TABLE 11.5 Descending Order of Flexural Strength Values for Ranked Resins

Flexural strength (kpsi) ASTM D-790	Resin	Rank
33.5	Polyethylene terephthalate (30% glass)	1
26.0–29.0	Polybutylene terephthalate (30% glass)	2
23.0	Polycarbonate (30% glass)	3
20.0–23.0	PPE-based resin (30% glass)	4
21.0 (est. value)	Nylon 6/6 (50% RH) (30% glass)	5
15.4	Polysulfone	6
13.5	Acetal	7
13.5	Polycarbonate	8
13.0	PPE-based resin	9
12.0	ABS	10
9.0	Nylon 6/6 (50% RH) (mineral reinforced)	11
9.0	Polypropylene (40% talc filled)	12
6.1	Nylon 6/6 (50% RH)	13
5.0	Nylon 6 (50% RH)	14
NA	Nylon 6/6 (50% RH) (Super Tough)	15

RANKING BY HEAT-DEFLECTION TEMPERATURE

In Chapter 3, on measurement of physical properties, the statement was made that heat-deflection temperature (HDT) does not usually indicate a practical use temperature, and therefore reference to it is discouraged. If it does have use at all, it is that it is more or less a measure of the temperature at which plastics have equivalent stiffness. Nevertheless, it has been used for years in the plastics industry as an empirical correlation to compare one resin to another in terms of physical response to temperature at a single load level (see Table 11.6). Of course, in recent times it has attained an upgraded status. It has been shown[48] that the single-point heat-deflection* temperature-versus-stress curves for different grades of a specific type of polymer as shown in Figures 3.9–3.11. Such a unification has a distinct advantage in that further experimentation is curtailed and heat-distortion curves for the grade of interest can be generated from the master curve through the single ASTM D-648 measurement.

RANKING BY TOUGHNESS (NOTCHED IZOD)

The notched Izod measurement of toughness of plastics is a test that leaves much to be desired. In addition to the notched Izod, there are a number of others, such as the tensile impact and direct impact tests, with none of them pro-

TABLE 11.6 Descending Order of Heat-Deflection Temperature (HDT) for Ranked Resins

HDT (° @ 264 psi) ASTM D-648	Resin	Rank
485	Nylon 6/6 (50% RH) (30% glass)	1
446	Nylon 6/6 (50% RH) (mineral reinforced)	2
435	Polyethylene terephthalate (30% glass)	3
405	Polybutylene terephthalate (30% glass)	4
345	Polysulfone	5
300	PPE-based resin (30% glass)	6
295	Polycarbonate (30% glass)	7
277	Acetal	8
270	Polycarbonate	9
265	PPE-based resin	10
216	ABS	11
194	Nylon 6/6 (50% RH)	12
170	Nylon 6 (50% RH)	13
170	Polypropylene (40% talc filled)	14
159	Nylon 6/6 (50% RH) (Super Tough)	15

*Value obtained under ASTM D-648 conditions can be used as a normalizing factor to coalesce heat deflection.

ducing satisfyingly reproducible results. Product literature prepared by Celanese[107] presents the situation realistically in the following way:

> The results of standard impact tests, such as the notched Izod test, frequently bear little relationship to the impact performance of molded parts under actual service conditions. It is not unusual for notched Izod results to rank materials in an entirely different order of merit than is warranted by their end-use performance. This lack of correlation exists because the test results are, in all cases, dependent upon such complex factors as part geometry and wall thickness, rate of loading, stress concentrations, and combinations of stresses. Nevertheless, the notched Izod test continues to be used as a reference for comparison of materials.

This is an excellent presentation of the situation and, from it, it can be deduced that small differences in notched Izod results from one resin cannot be expected to show an advantage or disadvantage. Nevertheless, the data are available and provide at least an order-of-magnitude idea of the difference in toughness within the group of resins being considered (see Table 11.7).

One further consideration about impact is that different polymers react differently to repeated impacts. For example, repeated tensile impacts at a level below normal break point for polycarbonate will cause breakage after a few

TABLE 11.7 Descending Order of Notched Izod Values for Ranked Resins

Notched Izod at 73°F (ft · lb/in.) ASTM D-256	Resin	Rank
40.0	Nylon 6/6 (50% RH) (Super Tough)	1
15.0	Polycarbonate	2
6.0	ABS	3
5.0	PPE-based resin	4
2.6	Nylon 6 (50% RH)	5
2.3	Nylon 6/6 (50% RH)	6
2.3	PPE-based resin (30% glass)	7
2.0	Polycarbonate (30% glass)	8
2.0	Nylon 6/6 (50% RH) (30% glass)	9
1.9	Polyethylene terephthalate (30% glass)	10
1.8	Polybutylene terephthalate (30% glass)	11
1.4	Acetal	12
1.4	Polysulfone	13
1.0	Nylon 6/6 (50% RH) (mineral reinforced)	14
0.4	Polypropylene (40% talc filled)	15

blows applied to that polymer, but nylons and acetals can withstand many blows at impact levels just below those that cause breaks in the tensile impact test. Unfortunately, literature is not readily available on this point for all the resins discussed here, nor is there any universally accepted test to illustrate repeated-impact affects.

RANKING BY ELONGATION AT BREAK

In the opinion of some people experienced in the use of engineering plastics, the elongation, or liability to "stretch" before breaking, is an indication of toughness; for those, its ranking is shown in Table 11.8. Measurement of elongation at break is normally obtained during the procedure for determining tensile strength, in which the speed used by the testing device to pull apart the test specimen is usually slow—often only 0.2 in./min, hardly an "impact."

On the other hand, the notched Izod method of measuring toughness, as has already been mentioned, has its limitations too, one of them being that some resins have greater notch sensitivity than others. Further, neither test gives a clue to the effects of repeated impacts. However, with the exception of the unreinforced nylons that tend to orient as they are pulled, leading to considerable elongation without necessarily indicating great toughness, there is a loose correlation between Izod and elongation. Normally, however, in the resin selection

Table 11.8 Descending Order of Values for Elongation at Break for Ranked Resins

Elongation at break (%) ASTM D-638	Resin	Rank
300	Nylon 6 (50% RH)	1
300	Nylon 6/6 (50% RH)	2
215	Nylon 6/6 (50% RH) (Super Tough)	3
110	Polycarbonate	4
75	Polysulfone	5
60	PPE-based resin	6
50	Acetal (avg. homopolymer and copolymer)	7
5–25	ABS	8
10	Nylon 6/6 (50% RH) (mineral reinforced)	9
8	Polypropylene (40% talc filled)	10
5	Nylon 6/6 (50% RH) (30% glass)	11
5	PPE-based resin (30% glass)	12
4	Polycarbonate (30% glass)	13
3	Polyethylene terephthalate (30% glass)	14
3	Polybutylene terephthalate (30% glass)	15

process, it is suggested that more emphasis be placed on the Izod in spite of its limitations.

RANKING BY APPARENT (CREEP) MODULUS

Creep is a very important property to consider in the process of selecting engineering plastics. It is defined simply in the *Modern Plastics Encyclopedia*[108] as follows:

> When a plastic is subjected to a constant static load, it deforms quickly to a strain roughly predicted by its stress/strain modulus and then continues to deform at a slower rate indefinitely or, if the load is high enough, until rupture occurs. This phenomenon, which also occurs in soft metals and structural metals at a very high temperatures, is called creep.

In designing with plastics, the projected life of a part, the end-use temperature, and the expected load on the part are all crucial to adequate design. Such data are generally obtained in terms of creep or stress relaxation variation with time ranging from a fraction of a second up to days and even years. The cost of experiments required for generation of stress/strain/time function up to an elapsed time of a few years is very high; an evaluation of one sample of a particular grade and at a particular temperature does not eliminate the need for data generation for other grades or for the same grade at different temperatures. Further, if data at a temperature other than that obtained in the experiments is required, one has to resort to bold interpolation or extrapolation or generation of new data, which is time-consuming and uneconomical. Thus it would certainly be attractive if a master curve could be evolved through careful analysis of existing data, so that a method emerges whereby tests of shorter duration could be carried out to predict long-term performance. Such a method is available,[47] and a bank of master curves can be generated along the lines discussed in Chapter 3. Through a careful behavioral pattern study of the stress-versus-time curves at different temperatures, a normalizing factor is deduced such that a single master curve can be formed through the use of the stress value at an elapsed time equivalent to 1 day, as shown in Figure 3.18–3.25. It is clear that the benefit of such a master curve is that long-term test programs can be curtailed to a day and reasonable estimates, which are good enough for design calculations, of the behavior in terms of long-term mechanical performance can be determined.

There are a number of variables to consider in obtaining creep data that can be used in these design calculations—so many, in fact, that the subject of creep could easily give rise to a book of its own. However, the point here is to acknowledge the importance of creep in resin selection, and to provide a ranking of the resins in terms of relative creep resistance.

To develop the ranking, it was necessary initially to obtain some quantitative measurement of creep of the plastics ranked here, and that is not very easy. Some material suppliers show creep rupture data, or apparent creep modulus, which are not the same thing; others report creep measured at one load level, still others report a different load level; and some do not report creep data at all. Fortunately, in the *Modern Plastics Encyclopedia*[108] apparent creep modulus for most of the 15 resins being compared here is reported, although not always at the same initial stress levels.

For this reason, it was necessary to calculate creep from the data presented in order to get all the resins on the same basis of temperature, load, and time. From this, as well as a few isolated independent sources of information, a ranking has been prepared that should indicate the order of creep resistance of the 15 resins (Table 11.9).

Obviously, as the resin selection process begins to narrow down the choice to a few viable candidates, accurate creep data on those resins should be obtained from the material suppliers. It is a rare part that can be adequately designed without creep calculations and, to make the calculation reliable, creep data must be obtained.

TABLE 11.9 Descending Order of Creep Modulus Values for Ranked Resins

Apparent (creep) modulus (kpsi)	Resin	Rank
1300	PPE-based resin (30% glass)	1
1000	Polyethylene terephthalate (30% glass)	2
750	Polybutylene terephthalate (30% glass)	3
730	Polycarbonate (30% glass)	4
600	Nylon 6/6 (50% RH) (30% glass)	5
450	Polypropylene (40% talc filled)	6
380	Nylon 6/6 (50% RH) (mineral reinforced)	7
350	Polysulfone	8
335	PPE-based resin	9
320	Polycarbonate	10
250	Acetal	11
215	ABS	12
83	Nylon 6/6 (50% RH)	13
63	Nylon 6 (50% RH)	14
39	Nylon 6/6 (50% RH) (Super Tough)	15

RANKING BY DIELECTRIC STRENGTH

The electrical properties of most plastics are excellent, particularly for those included in this comparison of engineering plastics. Since dielectric strength is a measure of the electrical strength of a material as an insulator, it is one of the two measurements of electrical properties to be important enough to be ranked here.

The dielectric strength is the dc voltage at which dielectric failure or breakdown occurs in the form of a continuous arc through the test specimen. The thinner the test specimen, the higher the reported dielectric strength, so reporting just the strength without the thickness is insufficient. Unfortunately, material suppliers have not been uniform over the years in selecting a common thickness of the specimens for dielectric strength. Most of the data in the following ranking table are values reported for ⅒-in.-thick samples, but some were not available at exactly that thickness and so their values have been estimated via interpolation or extrapolation. As a result, Table 11.10 can serve the useful purpose of presenting comparative values, but should not be used for design purposes where accuracy and dependability are required.

It should be kept in mind that the ranked resins are not compositions containing special stabilizers for continuous use at higher temperatures, or for UV stabilization, etc. If it becomes apparent that such compositions are needed, the material supplier should be contacted. The ranking, then, represents the perfor-

TABLE 11.10 Descending Order of Dielectric Strength Values for Ranked Resins

Dielectric strength (v/mil) ASTM D-149	Resin	Rank
550	Polyethylene terephthalate (30% glass)	1
550	PPE-based resin	2
550	PPE-based resin (30% glass)	3
500	Acetal	4
475	Polycarbonate (30% glass)	5
475	Polybutylene terephthalate (30% glass)	6
430	Polypropylene (40% talc filled)	7
430	Nylon 6/6 (50% RH) (mineral reinforced)	8
425	Polysulfone	9
420	Nylon 6/6 (50% RH) (30% glass)	10
420	Nylon 6/6 (50% RH)	11
420	ABS	12
400	Nylon 6 (50% RH)	13
380	Polycarbonate	14
380	Nylon 6/6 (50% RH) (Super Tough)	15

mance of only the base polymers or the base polymers plus glass or mineral reinforcing agents.

RANKING BY ARC RESISTANCE

One other electrical property that is important in designing for applications that might require good insulating characteristics is arc resistance. Test values for arc resistance vary considerably with specimen thickness, but in principle, arc resistance is a measure of the ability of a material to resist the action of a high-voltage, low-current dc arc.

In some polymers, such as acetals, the reported arc resistance value is not very producible because the polymer tends to melt, or even burn, under test conditions and thus obscures a precise value.

With one or two exceptions, the values in Table 11.11 are those listed by Underwriters Laboratory and, of course, represent the performance of the base polymers without additives, except for glass and mineral reinforcement.

RANKING BY SPECIFIC GRAVITY

Not much needs to be said about specific gravity testing procedures—they are accurate. Specific gravity per se is not often a criterion for resin selection.

TABLE 11.11 Descending Order of Arc Resistance Values for Ranked Resins

Arc resistance (s) ASTM D-495	Resin	Rank
240	Acetal	1
146	Polybutylene terephthalate (30% glass)	2
142	Polypropylene (40% talc filled)	3
139	Nylon 6/6 (50% RH) (mineral reinforced)	4
135	Polyethylene terephthalate (30% glass)	5
131	Nylon 6/6 (50% RH) (Super Tough)	6
120	Polysulfone	7
116	Nylon 6/6 (50% RH)	8
96	Nylon 6/6 (50% RH) (30% glass)	9
83	ABS	10
75	PPE-based resin	11
65 (est.)	Nylon 6 (50% RH)	12
10	Polycarbonate	13
10	Polycarbonate (30% glass)	14
9	PPE-based resin (30% glass)	15

There are occasions when the weight of a part becomes important, however, and the rankings shown in Table 11.12 simplify the process of making simple weight comparisons between the resins.

RANKING BY TYPICAL MOLD SHRINKAGE (AND EFFECTS ON WARPAGE)

A number of factors affect the extent of shrinkage of engineering plastics. Inherently, crystalline polymers such as acetals and nylons shrink more than amorphous polymers such as polycarbonate and PPE-based resins. In addition, mold shrinkage depends on part thickness, geometry, and processing conditions. Finally, the amount of reinforcing agent or filler in a resin affects the extent of shrinkage of that resin and also affects the direction of the shrinkage in the part.

Since so many variables enter into part shrinkage, it is not simple to rank it. Nevertheless, using "typical" reported values, the ranking in Table 11.13 can serve as a guide, although the values cannot be used for mold design purposes.

It would be helpful if expected warpage correlated with predicted shrinkage, but this is usually not the case. To quote from an article in *Plastics Technology* by Peter J. Cloud and Mark A. Wolverton:[109]

TABLE 11.12 Ascending Order of Specific Gravity Values for Ranked Resins

Specific gravity ASTM D-792	Resin	Rank
1.05	ABS	1
1.06	PPE-based resin	2
1.10	Nylon 6/6 (50% RH) (Super Tough)	3
1.13	Nylon 6/6 (50% RH)	4
1.13	Nylon 6 (50% RH)	5
1.20	Polycarbonate	6
1.22	Polypropylene (40% talc filled)	7
1.24	Polysulfone	8
1.27	PPE-based resin (30% glass)	9
1.30	Nylon 6/6 (50% RH) (30% glass)	10
1.40	Acetal	11
1.43	Polycarbonate (30% glass)	12
1.51	Nylon 6/6 (50% RH) (mineral reinforced)	13
1.53	Polybutylene terephthalate (30% glass)	14
1.56	Polyethylene terephthalate (30% glass)	15

TABLE 11.13 Ascending Order of Approximate Shrinkage Values for Ranked Resins

Shrinkage (mils/in.) ASTM D-955	Resin	Rank
(1–3)	PPE-based resin (30% glass)	1
(2–3)	Polycarbonate (30% glass)	2
(4–8)	Polyethylene terephthalate (30% glass)	3
(4–6)	Polybutylene terephthalate (30% glass)	4
(5–7)	PPE-based resin	5
(5–7)	Polycarbonate	6
(6–10)	Nylon 6/6 (50% RH) (30% glass)	7
7	Polysulfone	8
8	ABS	9
9	Nylon 6/6 (50% RH) (mineral reinforced)	10
10	Polypropylene (40% talc filled)	11
15–20	Nylon 6/6 (50% RH)	12
15–20	Nylon 6/6 (50% RH) (Super Tough)	13
15–20	Nylon 6 (50% RH)	14
22	Acetal	15

All too often, shrinkage values listed on data sheets are thought to indicate probable part warpage [as well]. This information typically derives from 5 in. × ½ in. × ⅛ in. rectangular bars measured in accordance with ASTM D-955 test methods. These specimens, however, ignore the anisotropic shrinkage characteristics of resins containing fiber reinforcements and/or particulate fillers, and are used to report shrinkage in one direction only—along the length of the bar. As a result, there is little correlation between shrinkage values and warpage factor.

It is interesting to note that, in general, for crystalline glass-reinforced polymers such as nylons, optimum properties of stiffness, creep, and tensile strength are obtained at the expense of increased warpage, although the shrinkage is less. This is true of warpage attributable to material composition, which can affect the warpage drastically if it is not properly considered in the part design. It is well to keep in mind that for thin parts, glass-reinforced resins warp more as glass loading increases. Paradoxically, heavy section parts in glass-reinforced resins warp less as the glass loading increases.

RANKING BY PRICE

Obviously, the price of the resin to be selected is a very important "quality." To consider the price only in terms of cost per pound is not so important as to consider its cost in relation to its volumetric cost. In Table 11.14 are listed the prices of the 15 resins in terms of cost per cubic inch.

Prices fluctuate in part as dictated by worldwide economic conditions. The prices shown here were taken from the March 1996 issue of *Plastics Technology*[105] and are based on truckload quantities. As is the case for most of these rankings, the values are not absolute but are suitable for providing comparative costs.

RANKING BY PROCESSIBILITY

It would be ideal to be able to rank resins in the order of the ease of processibility, but a judgment of that property does not lend itself to substantiating data. For this reason, a ranking of the property is not provided here. However, to provide some guidance, albeit slight, a few observations follow:

1. Glass- or mineral-reinforced polymers are not significantly difficult to mold, but they present more problems than their base resins. Largely the problems include the erosive action of the reinforcing agents, special attention to molding conditions to produce a reasonable surface on the molded parts, and usually (but not always) more problems with warpage.

2. The unreinforced resins in the ranking will be the easiest to process; but even with this group, some will obviously mold a little more easily than

TABLE 11.14 Ascending Order of Price in Cents per Cubic Inch for Ranked Resins

Price (¢/in.^{3})[a]	Resin	Rank
2.3	Polypropylene (40% talc)	1
3.4	ABS	2
5.2	PPE-based resin	3
5.9	Nylon 6 (50% RH)	4
6.3	Acetal	5
6.3	Nylon 6/6 (50% RH) (mineral reinforced)	6
6.5	Nylon 6/6 (50% RH)	7
6.7	Polycarbonate	8
7.4	Polyethylene terephthalate (30% glass)	9
7.7	Nylon 6/6 (50% RH) (Super Tough)	10
9.8	Nylon 6/6 (50% RH) (30% glass)	11
10.0	Polybutylene terephthalate (30% glass)	12
10.4	Polycarbonate (30% glass)	13
10.7	PPE-based resin (30% glass)	14
20.9	Polysulfone	15

[a]*Source*: Ref. 105.

others. Most of them require attention to water content before molding, and desiccant-type dryers should be used for effective drying with a minimum of time and cost.

A rather subjective ranking of processibility of this group of resins would result in the following order according to ease of processing:

Nylon 6
Nylon 6/6
Nylon 6/6 (Super Tough)
Acetal
ABS
PPE-based resin
Polycarbonate
Polysulfone

EXAMPLES OF SELECTION

Example 1

The selection process goes on every day in industry, of course, and many examples of the selection process could be envisioned. In this example, let us assume that end users unfamiliar with engineering plastics decide that a trigger for a special metal hand drill already on the market should be replaced with a suitable plastic to reduce production costs. This might be a step toward replacing more of the metal parts of the drill.

The end users are obviously familiar with what the part must do in use, and have a reasonable feel for what some of the material properties must be to make the use successful. Let us further assume that they decide to refer to earlier parts of this chapter to get some preliminary ideas about which engineering plastics reviewed should receive further consideration. If they have other candidates they want to include in their preliminary comparison of compositions, they can simply obtain the appropriate physical property data and include them in the ranking process.

They could then proceed in the following manner.

General information

1. Part under consideration is a trigger for a hand drill.
2. Cost must be reasonable for the plastic, and the part finished cost is significantly less than that of the current metal part.
3. The plastic part must be used without any design changes being made to the hand drill itself.
4. In this very early stage of development, little quantitative design data have been prepared.

Codes and specifications

Their device is not covered by any industrial or military codes or specifications.

Cost

Cost is important, since this is the major motivation for the switch to plastic.

Environment

The hand drill is frequently used in specialized industrial conditions, including exposure to a number of aromatic organic solvents, ketones, and aldehydes. The drill is occasionally exposed to high temperatures for brief periods of time; and, for design purposes, the end users want the trigger to have a heat-deflection temperature of at least 250°F.

Structural considerations

Qualitatively, the end users know they need a plastic material with high stiffness, good resistance to creep, and reasonable toughness. Their intent is to narrow down the selection qualitatively and then get into design calculations in depth.

Other design considerations

The end users expect the proposed part to be wear resistant; and because it is a straight replacement of a metal part in a piece of machinery, they need parts molded to close tolerances that will retain reasonable dimensional stability. Esthetics are not critical, but a specific color will be needed and the surface should be uniform, although not necessarily glossy.

With these few criteria, the end users can proceed to make a preliminary elimination of those compositions that would not be adequate for their application and create a shorter list of those that might be adequate. A simple approach would be to lay out a form such as shown in Table 11.15, which includes columns for the 15 resins (or others, if they want to include some) along with key properties to be considered. By referring to the various ranking tables, they can then begin to fill in the adequacy of the various compositions. Obviously, it would be better if quantitative objectives for the key properties were available, but at this point it is assumed that they are not; so estimates have to be made to make some progress in the process.

To illustrate how the adequacies for each property were estimated for this example, each consideration is reviewed below.

Price. By reviewing the price ranking in Table 11.14, the end user can eliminate polysulfone, since it is at least twice as expensive than most of the other candidates in cents per cubic inch. Later in the selection process, when other factors have been considered and some candidates have been selected, the costs can be considered more on a comparative basis.

TABLE 11.15 Worksheet for Example 1

Plastic composition	Heat-deflection temperature	Stiffness	Cost	Environment	Toughness	Creep	Tensile strength
Polyethylene terephthalate (30% glass)	OK	OK	OK	No	OK	OK	OK
Polybutylene terephthalate (30% glass)	OK	OK	OK	Maybe	OK	OK	OK
PPE-based resin (30% glass)	OK	OK	OK	No	OK	OK	OK
Polycarbonate (30% glass)	OK	OK	OK	No	OK	OK	OK
Nylon 6/6 (30% glass)	OK	OK	OK	OK	OK	OK	OK
Nylon 6/6 (mineral reinforced)	OK	No	OK	OK	No	OK	OK
Polypropylene (40% talc filled)	No	No	OK	No	No	OK	No
Acetal	OK	No	OK	OK	Maybe	Maybe	OK
Polysulfone	OK	No	No	No	Maybe	Maybe	OK
ABS	No	No	OK	No	OK	No	No
PPE-based resin	OK	No	OK	No	OK	Maybe	OK
Polycarbonate	OK	No	OK	No	OK	Maybe	OK
Nylon 6/6	No	No	OK	OK	OK	No	OK
Nylon 6	No	No	OK	OK	OK	No	OK
Nylon 6/6 (Super Tough)	No	No	OK	OK	OK	No	OK

Environment. The lack of adequate resistance to crazing or attack by aromatic organics, aldehydes, and ketones (see Table 11.2) will usually preclude the use of Noryl compositions, polyethylene terephthalates, or polycarbonates.

Structural. Assuming the quantitative modulus has not yet been determined, by reviewing the ranking of modulus it can be seen that the first five resins listed, all glass reinforced, are significantly stiffer than the remainder of the group, (see Table 11.4) and can be considered strong candidates for the proposed application.

The end users in this example have decided that a heat-deflection temperature of 250°F or higher is necessary, which simplifies the selection process based on that particular property (see Table 11.6). Referring to Table 11.6, it can be seen that ABS, unmodified nylon 6/6 at 50% RH, toughened nylon at 50% RH, nylon 6 at 50% RH, and polypropylene (with 40% talc) can be eliminated from consideration.

Tensile strength, normally an important property to the designer, is not particularly critical in this example, although any composition with tensiles less than 5 kpsi is not normally considered for most engineering plastic applications. Therefore, only ABS and polypropylene (with 40% talc) would be eliminated on the basis of this property alone (see Table 11.3).

In reviewing the creep modulus of the resins, it is clear that the glass- and mineral-reinforced resins are superior to the others, but only those candidates with values lower than that for ABS have been arbitrarily considered inadequate for the proposed part (see Table 11.9).

Except for the extremes, the Izod measure of toughness is not particularly discriminating, as has been mentioned previously. Actually, a case can be made arbitrarily, based on practical end-use experience, that an Izod over 1.5 represents a resin that is reasonably tough. This decision would result in the elimination of acetals, mineral-reinforced nylon, and talc-filled polypropylene from consideration (see Table 11.7).

Conclusion

In this rather primitive example, and assuming the end users did not introduce candidates of their own into the rankings, a review of Table 11.15 indicates that, of the 15 ranked resins, two of them, glass-reinforced nylon 6/6 and glass-reinforced polybutylene terephthalate, have the most desirable properties and should be followed up in more detail. The end users could then begin to define quantitative design requirements and to use the physical property data of these two compositions, as well as cost calculations, to determine which will become the ultimate selection.

Before finalizing the choice, the end users should contact the material suppliers of these two compositions and consult in depth with them for confirmation of the decision or, possibly, to determine if even more appropriate versions of these compositions are available.

Example 2

To set up a second example of the use of the rankings in a selection process, let us assume the end users are interested in the manufacture of a housing of moderate size (6 × 12-in. surface area, 3 in. deep) for an electronic communications device. In this example, some preliminary calculations have been made, which is normally the case, and design data are available. It must be remembered that toughness, stiffness, tensiles, etc., requirements can be a function of part thickness and shape, among other considerations; and in this example the designer has presumably taken these facts into consideration in arriving at preliminary design criteria.

General information

1. The part is to be a housing for an electronic communications device.
2. The surface of the part must be attractive, with a uniform texture.
3. Close molding tolerances must be maintained.
4. The part must remain dimensionally stable in a normal ambient environment.

Codes and specifications

There are no specific code requirements.

Environment

1. The part is to be used in normal ambient conditions.
2. There will be no exposure to solvents or chemicals.
3. Internal temperature of the part will not exceed 250°F.

Cost

Within reason, the price of the proposed plastic is not a major consideration.

Structural considerations

The end user has arrived at some tentative requirements for the proposed plastic:

1. Good tensiles are needed—not less than 7 kpsi.
2. Although some latitude is available through design, the minimum modulus should be 300 kpsi.

3. The part must be able to stand some abuse in use, and the designer has set a minimum of 3.0 for an Izod.
4. Resistance to creep for the part is important, with a minimum creep modulus of 250 kpsi set by the designer.
5. The heat-deflection temperature must be at least 250°F.
6. Fit is very important in the proposed application, and the designer has established an arbitrary limit of 8 mils/in. in any direction for the mold shrinkage of the resin.
7. To function properly, the housing must be dimensionally stable, and it must not be affected by changes in ambient humidity.

Other design considerations

Esthetic considerations are important. The part surface must be uniform and attractive.

Conclusion

In this simplified example, having been rated "OK" on all the listed properties, it would appear from observation of Table 11.16 that only unmodified polycarbonate and unmodified PPE-based resin emerge as the primary candidates for further in-depth review.

Certainly, a cost estimate would be a part of the follow-up steps, where cost per cubic inch, molding cycles, optimum wall thicknesses, etc., can be examined and the two candidates compared on those bases.

One of the major limitations of this preliminary selection procedure is that it cannot take into consideration all of the many resin modifications available today. As just one example, both glass-reinforced polyethylene and polybutylene terephthalate in this ranking procedure example could have been selected as candidates for further study except for their apparent lack of sufficient toughness, yet there are commercial varieties of both of these polymers that are indeed toughened enough to attain an Izod higher than 3.0.

As a rule of thumb, it is suggested that when a resin being evaluated in the ranking comparison seems to fail in only one requirement, the material suppliers of such a composition should be consulted to see if the proper modification is available. If one is, it should become one of the "finalists" in the ranking comparison.

"GENERAL USEFULNESS" RANKING

There is, of course, no accurate way of proving that any resin is "better" than the others, since all 15 have attractive properties and are used in large volumes for a broad spectrum of end uses. Nevertheless, it is interesting to try to com-

TABLE 11.16 Worksheet for Example 2

Plastic composition	Heat-deflection temperature	Stiffness	Toughness	Creep	Tensile strength	Shrinkage	Dimensional stability
Polyethylene terephthalate (30% glass)	OK	OK	No	OK	OK	OK	OK
Polybutylene terephthalate (30% glass)	OK	OK	No	OK	OK	OK	OK
PPE-based resin (30% glass)	OK	OK	No	OK	OK	OK	OK
Polycarbonate (30% glass)	OK	OK	No	OK	OK	OK	OK
Nylon 6/6 (30% glass)	OK	OK	No	OK	OK	No	OK
Nylon 6/6 (mineral reinforced)	OK	OK	No	OK	OK	No	OK
Polypropylene (40% talc filled)	No	OK	No	No	No	No	OK
Acetal	OK	OK	No	OK	OK	No	OK
Polysulfone	OK	OK	No	OK	OK	No	OK
ABS	No	OK	OK	No	No	No	OK
PPE-based resin	OK	OK	OK	OK	OK	OK	OK
Polycarbonate	OK	OK	OK	OK	OK	OK	OK
Nylon 6/6	No	No	No	No	OK	No	No
Nylon 6	No	No	No	No	OK	No	No
Nylon 6/6 (Super Tough)	No	No	OK	No	OK	No	No

TABLE 11.17 "General Usefulness" Ranking

	PET Polyester (30% glass)	PBT Polyester (30% glass)	Nylon 6/6 (30% glass)	Acetal	PPE-based resin	Nylon 6/6 (mineral reinf.)	PPE-based resin (30% glass)	Polycarbonate (30% glass)
Tensile strength	1	4	2	8	9	10	5	3
Flexural modulus	1	2	5	8	11	6	3	4
Heat-deflection temperature	3	4	1	8	10	2	6	7
Notched Izod	10	11	9	12	4	14	7	8
Elongation	14	15	11	7	6	9	12	13
Creep	2	3	5	11	9	7	1	4
Dielectric strength	1	6	10	4	2	8	3	5
Arc resistance	5	2	9	1	11	4	15	14
Cost (¢/in.^{3})	9	12	11	5	3	6	14	13
	46	59	63	64	65	66	66	71
General usefulness ranking	1	2	3	4	5	6	7	8

TABLE 11.17 *(Continued)*

	Nylon 6/6	Polypropylene (40% talc)	Polysulfone	ABS	Polycarbonate	Nylon 6	Nylon 6/6 (Super Tough)
Tensile strength	6	15	7	14	11	12	13
Flexural modulus	13	7	9	10	12	14	15
Heat-deflection temperature	12	14	5	11	9	13	15
Notched Izod	6	15	13	3	2	5	1
Elongation	2	10	5	8	4	1	3
Creep	13	6	8	12	10	14	15
Dielectric strength	11	7	9	12	14	13	15
Arc resistance	8	3	7	10	13	12	6
Cost (¢/in.^{3})	7	1	15	2	8	4	10
	78	78	78	82	84	88	93
General usefulness ranking	9	10	11	12	13	14	15

pare all 15 in a single summary to determine if a ranking of "general usefulness" can be determined based on some physical properties alone.

Tables 11.3, 11.4, 11.6, 11.11, and 11.14 present quantitative values for the resins in descending order for the various properties. From these, a summary has been prepared as shown in Table 11.17. The values for shrinkage, specific gravity, and melting point were excluded because it was felt these properties were not "miscible" with the others in this particular kind of a summary. The method of construction of the summary is simple. Numbers 1–15 are assigned to the resins according to the optimum value of a resin for each of the properties listed. The assigned numbers are not weighted according to any arbitrary estimate of the degree of importance.

For example, under tensile strength, the value 1 for 30% glass-reinforced polyethylene terephthalate (PET) means that PET has the highest value, while the number 15 for 40% talc-filled polypropylene indicates it has the lowest tensiles of the resins being compared. In the case of toughness, Super Tough nylon 6/6, with a value of 1, has the highest or most desirable toughness according to the notched Izod test, while 40% talc-filled polypropylene, with a value of 15, has the lowest. As a final example, 30% glass-reinforced PPE-based resin has the best or most desirable creep properties, and Super Tough has the least.

The "total" of the assigned values is simply the empirical result of adding the values for each resin. It is stressed that these totals are empirical numbers, with no scientific justification. Nevertheless, they tend to rank the way the compositions line up when the specified physical properties are considered.

From the data in Table 11.17, several observations can be made.

1. It appears that 30% glass-reinforced PET has the best blend of properties compared to the others, with only toughness being marginal.
2. In terms of mechanical properties alone, the reinforced compositions are better than their base compositions and should be considered first when such properties are important in the selection process.
3. In general, the reinforced compositions have the poorest impact resistance.
4. In general, the tougher compositions have poorer mechanical properties.
5. Of the unreinforced compositions, PPE-based resin, acetal, nylon 6/6, polysulfone, and polycarbonate, in that order, offer the best "average" properties (and are easier to process).
6. There is no good correlation between cost in terms of cents per cubic inch and the "general usefulness" of a resin.

It may be that resin selectors can get a general impression of where to start in their selection process from this general usefulness exercise, but it is preferable to work from the individual quantitative ranking tables.

12

Reselection of Resin

INTRODUCTION

In a small but still surprising number of cases, end users make a decision to abandon the first resin of their choice for another. There is a variety of reasons for this.

Perhaps the first reason is that commercial experience with the item indicates the original resin selected was not adequate. It could be that an annoying number of failures in the marketplace have led to recalls, with accompanying bad press and expense. It could also be that, as a result of price shifts, the original choice is more expensive than other possibilities.

Another reason for wanting to change is that other items in an end user's line of products may be all in one other resin, and, for the sake of economics and simplicity, end users may want to put all of their products in the same material, if possible.

Occasionally, a product line, or an entire business, is bought out by a new owner, who then makes a decision that the original resin selection should at least be reviewed. Or it has happened that end users have moved their molds to new molders who begin to agitate for a switch to another resin, either because they think it is a better choice or because they do not really like to mold the original resin.

Actually, all of these examples, with the exception of field failures, are reasonably commonplace. The end result is that the process of resin selection must be gone over again, but this time there are a few new restrictions, since the expensive mold has already been made.

CONSIDERATIONS

The selection process now has new facets to it, since the parameters of the mold already in hand eliminate certain degrees of flexibility. Some of the problems to consider when reviewing the options of replacing one resin with another when the mold is already in hand are as follows.

Shrinkage

Serendipity is required to find a resin in one family of polymers that will have the same shrinkage as one in another family. For example, it is difficult to replace a low-shrink amorphous resin with a high-shrink crystalline one. Certainly, an unfilled or unreinforced nylon will shrink significantly more than an unfilled or unreinforced polycarbonate. If dimensions are a critical consideration, a switch between these two polymers is ruled out. On the other hand, some of the nylons, glass reinforced or mineral filled, have roughly the same shrinkage as the natural polycarbonate, in which case, a switch is feasible based on shrinkage alone.

Warpage

The amount of warpage, if any, is likely to vary from one resin to another, although it is scarcely predictable. Tinkering with molding conditions can sometimes mitigate the extent of warpage, but if the end user demands absolute freedom from warpage, and the original resin selection was heavily weighted on that property, it will be fortuitous indeed if a secondary selection is as warp free. Of course, much of this depends on the geometry of the part. Only a trial run of the secondary resin in the original mold will supply the answer. Adjusting molding conditions may mitigate the degree of warpage to some extent, but the end user may have to accept something less than ideal.

Cycle

It is likely that a change in resin will result in a change in the molding cycle. Of course, it is always possible that the second choice will have a faster cycle, which would, of course, be an advantage. On the other hand, the cycle could be longer and could be even so much longer, so that any advantage anticipated from a lower price may be offset by the new cycle. It is not a simple thing to predict accurately what difference in cycle will be between two resins, but with the mold in hand, it is easy to check it out on the molding machine.

Surface

The as-molded surface of parts can vary considerably from one polymer to another, and the selection process for the second-choice resin should include what is known on the subject. For example, a heavily loaded glass-reinforced resin is not going to have the gloss of a resin without glass; and if appearance is important, the selection can be limited to the natural resins. Or, in a common marketing ploy, surfaces of a mold can be finished to provide the parts with a matte finish, which may obscure major surface differences between natural- and glass-reinforced resins—thus permitting interchangeability between them from the standpoint of both shrinkage and surface.

When only a natural resin can be used to replace another with different shrinkage values, flexibility in switching is considerably reduced; or it can be eliminated altogether. In general, nylons and polycarbonates, for example, can be molded into parts with nice glossy surfaces, but if mold shrinkage is important, it is unlikely that the switch can be made at all.

Flash

Flash is a rather minor consideration, but it is possible that a mold that is satisfactory for a low-melt-flow resin, such as polycarbonate, would flash noticeably when a high-melt-flow resin, such as a nylon 6/6, is molded in it. This can only be verified by an actual molding run.

13

Effects of Recycling on Resin Selection

INTRODUCTION

The volume of plastics being used today is huge and growing. In addition, new plastic resins and uses are developing rapidly, as value-in-use is constantly being upgraded through improved and innovative design. Ultimately, however, the time comes when the useful life of products made from these resins end and they must be disposed of. At the same time, public concern continues to grow about the large volumes of waste materials, including plastics, that are already being landfilled or burned, contributing to an already alarming growth in air, water, and land pollution. It is apparent that immediate steps must be taken to reduce the volume of plastics involved in waste disposal problems without restricting the essential growth in the volume used. Any reduction in plastics consumption is to be considered neither progress nor an economic saving. The materials that would be used in their place would not necessarily be more economical. Our main concern should rather be recycling of plastics wastes, thereby enabling their energy value to be maintained. The recovery of economic value from plastics waste is a well-treated subject,[110,111] and various methods of doing so are available.

To appreciate the critical need for increased and improved recycling, it is helpful to understand the mounting crisis of waste disposal. In 1990 approximately 195.7 million tons of municipal solid waste (MSW) were generated. About 67% of this went into approximately 5000 municipal landfills. Some of these landfills are rapidly approaching capacity, and unless something is done to reduce the MSW volume, new sites will have to be developed, and expensive delivery costs to more remote areas can be expected. Landfilling is already an expensive operation, without this added cost. For example, in the 1970s the landfill cost per ton was only a few dollars, while today in some places in the United States the cost is over $100 per ton.

About 16% of the 197.5 million tons of MSW generated in 1990 was incinerated, and the remaining 17% was recycled or composted. Most of the recycled material was paper and consumer plastics used in packaging. Recycling of nearly a billion pounds of consumer plastics alone reflects real progress, but more is needed. In addition to municipal landfill areas, plastics, particularly the durable plastics, are finding their way into industrial landfills. In terms of durable plastics, recycling is just beginning, with the impetus coming from automobile and plastics producers.

Obviously, public concern about landfill has added urgency to the need to minimize the volume of waste, and recycling and source reduction are obvious approaches. The recycling of plastics has progressed to the point that beverage containers are being made back into beverage containers and other food-contact containers. Milk, juice, and water containers are reused in production of household chemical bottles, and various other plastics are recycled into other applications, such as garbage cans, lumber, and office products. A novel concept of recycling thermoplastic polyester polymers (PET) from used beverage bottles into raw materials for use in manufacturing of unsaturated polyester resins has been presented in the past.[112] Such innovative approaches are continually being sought.

Public concern carries over to government, which has already considered the possibility of restricting the amount of plastics that can be used, or in some cases eliminating certain types of plastic packaging altogether. For years, plastics processors have been attempting to make polymers resistant to UV light degradation through the addition of carbon black or chemical UV inhibitors. However, with the advent of litter legislation to clean up the environment, the government, at the urging of environmentalists, has considered legislation to require the plastic rings connecting beverage cans to be either bio-, photo-, or chemically degradable. Since spontaneous chemical degradation does not seem feasible and attempts to make plastics truly biodegradable have not been successful, photodegradation has been explored.[38,113] Environmental concerns are giving rise to a lot of new research. Improving the environmental impact of plastics will go a long way in reducing the search for new and improved recycling techniques.

Another very important recycling incentive is economics. In contrast to the concern about increasing volumes of MSW destined for landfills is the need for plastics in large appliances, building and construction materials, business machines, computers, and automobiles. This is particularly true in the automotive industry, where the use of durable plastics enhances fuel economy by making cars lighter and reducing energy requirements during the life of the vehicle.

Adding impetus to the need for improved recycling of durable plastics is the work being done in France,[114] Germany,[115] the Scandinavian countries,[116] the United Kingdom, Japan, and the United States[117]. It is in the best competitive interests to devote more money, time, organization, and effort to maintaining a respectable position.

AMERICAN PLASTICS COUNCIL

The organization leading the United States today in the search for optimum recycling methods and procedures for durable plastics is the American Plastics Council (APC). "Durable" plastics are defined by the APC as ones that have at least a 3-year useful life, and include the usual engineering plastics such as nylons, acetals, polycarbonates, etc., as well as polyvinyl chloride, sheet molding compound (SMC), some fiberglass-reinforced polyesters, and a number of thermosets. However, for the purpose of discussion in this book, "durable" plastics will be limited to the engineering thermoplastics.

The APC was created in 1991 as a result of recognition by the plastics industry that it needed to remain ahead of the curve on the issue of plastics use in durable applications. The Durables Committee of the APC was formed and is funded currently by the 27 member companies of the APC and is located in Washington, D.C. The APC is a joint initiative with the Society of the Plastics Industry. The committee focuses on the following market segments:

Automobiles
Major appliances
Building and construction
Furniture

Consumer electronics
Small household appliances

The objectives of the APC include leadership in locating, developing, and promoting technology to address sound recycling and/or disposal of durable plastics, including energy recovery when that is in order. The APC accomplishes these objectives by supporting research and development, and by initiating and following up on studies. A Durables Committee task force of the APC has been engaged in this activity since 1990.

THE AUTOMOTIVE INDUSTRY

One of the largest users of durables, including engineering thermoplastics, is the automotive industry. The growth of plastics in this industry has been quite rapid, particularly over the past 10 years. There are a number of reasons for this rapid growth. Improvement in processing technology in injection molding, blow molding, and thermoforming are partly responsible. In addition, plastic resins have been developed that deliver high levels of performance in various applications. In addition, safety considerations have played an important part. Certainly the airbag is an excellent example. Other examples of safety improvements are safer gas tanks, seat belts, and safety glass inner layers.

Certain refinements in under-the-hood parts have been made possible through the use of plastics. Further, in automobile design, the use of plastics results in flexibility of the car design concept, improved quality, dent resistance of some body parts, and weight reduction. As mentioned previously, weight reduction is particularly important, since it results in energy savings by reducing the amount of gas the car uses. In fact, Franklin Associates, of Prairie Village, Kansas, have estimated that vehicles made in 1988 would save 112 trillion BTUs over their lifetime, the equivalent of 21 million barrels of oil saved due to decreased weight provided by plastics. As a result of the increased use of plastics in cars, about 8% of a car's weight currently is plastic, or about 300–400 lb.

Of course, the automobile has a finite life, and when it is declared ready to be retired and scrapped, a waste disposal challenge arises. Presently, 90% of all junked cars are recycled, and reuse of them usually follows a fairly routine procedure. Reusable parts are removed by the scrap dealer for resale as is—and a few of these parts are made from plastics. The stripped car is placed in a field for about a year as a possible source of specifically and seldom-requested parts. From the field the car is removed to shredders, who put it through a hammer mill and chop it into fist-sized pieces of metals, glass, rubber, plastics, etc.

Steel is removed from the chopped material via magnets, and any light fluffy material is blown off and collected. The remainder of the chopped material contains nonferrous metals, plus heavier pieces of plastic, rubber, and glass. This mixture is sold to a nonferrous separator, who removes the nonferrous metal by a special process, leaving a residue of glass, rubber, dirt, some metal, and plastic. Plastic accounts for about 25% by weight of this mixture. The automotive shredder residue (ASR) is typically landfilled, usually in industrial landfills, since further separation is not feasible at the present time. This same general procedure applies to some major appliances such as refrigerators, washing machines, dryers, etc.

A simple example of the recycling of durables originates in the computer industry. Unlike the automotive problems of obtaining clean material for recycling, computers being junked, usually for reasons of obsolescense, can readily be isolated into housings and internals. With care and close attention to cleanliness, the housings can be reground and remolded into other articles.

Although separation of plastics from the shredded residue of the nonferrous separation process in the automotive example is not practical, chemical companies are examining ways to extract some of the energy and/or chemical building blocks in the plastics. Two methods include the processing of the residue by hydrolysis or alcoholysis (see Figure 13.1). In these methods, superheated steam or alcohol, depending on the general nature of the polymer in the residue, is used to depolymerize the plastic into its original building blocks. Chemical companies already have experience with this depolymerization process. Emphasis in the industry is currently being placed on the recovery of materials from the shredded residue, which contains blends, alloys, stabilizers, reinforcements, fillers, tougheners, and contaminants.

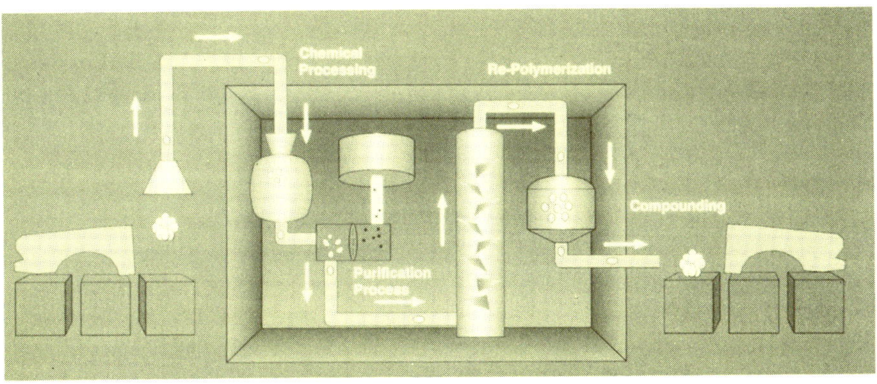

FIGURE 13.1 Schematic of hydrolysis and alcoholysis process. (Courtesy of American Plastics Council, Washington, D. C.)

FIGURE 13.2 The tall ship HMS *Rose* with 13,000 square feet of sails produced from recycled polyester. (Courtesy of E. I. du Pont de Nemours and Co., Wilmington, DE.)

Figure 13.2 is a dramatic example of what can be accomplished via depolymerization and repolymerization of polyester. The 13,000 ft^2 of sails were recycled from soda bottles (see Figure 13.3) and recently introduced plastic car fenders (see Figure 13.4). E. I. du Pont de Nemours used a proprietary depolymerization process to break down the shredded car fenders into basic building blocks, removed impurities, and created recycled resin from the building blocks. In addition, some 126,000 used polyester soda bottles were recycled through conventional mechanical processes. The resin supplied from these two sources together were extruded into yarn, from which the sails were made.

Du Pont has firm plans to expand the alcoholysis approach to recycling of polyester as used in the example above to re-create the resin from the building blocks. To demonstrate this resolve the company has started up a 15-million-pound-per-year methanolysis plant at a company site in Nashville.

To carry the recycling effort still further, emphasis is currently being placed on the recovery of materials from the shredded residue from junked cars, which contains blends, alloys, stabilizers, reinforcements, fillers, tougheners, and contaminants.

FIGURE 13.3 More than 126,000 plastic soda bottles and recycling technology were used to help produce the sails for the HMS *Rose*. (Courtesy of E. I. du Pont de Nemours and Co., Wilmington, DE.)

In addition to hydrolysis and methanolysis, there is another method of depolymerization. This is pyrolysis, where the plastics are heated under pressure in the absence of oxygen, and broken down into feed streams which may include some monomer. The predominate products include oil and gases. The gas can be used as fuel for the heating process.

The APC, along with the automotive and plastics industries, is devoting significant effort to improving plastics recycling techniques. For the present and immediate future, this effort will be concentrated on two approaches. One of these is to improve current methods of recovering plastics from scrapped cars through selective dismantling, and the other is to concentrate on the design for disassembly of cars now on the drawing board. Those with the responsibility for selecting thermoplastics for engineering applications should be aware of this.

SELECTIVE DISMANTLING

One ongoing and relatively simple way to reduce the percentage of engineering plastics from cars now being landfilled is selective dismantling. Dismantlers have begun to remove plastic parts from junked cars, including under-the-hood

FIGURE 13.4 An example of a polyester-based fender that was used in recycled form to make yarn for the sails of the HMS *Rose*. (Courtesy of E. I. du Pont de Nemours and Co., Wilmington, DE.)

FIGURE 13.5 A radiator endcap being removed for recycling. (Courtesy of American Plastics Council, Washington, D. C.)

parts, and to collect them in containers (see Figures 13.5 and 13.6). A secondary selective process then separates these parts into appropriate families of resins, to be dealt with later. These actions are taken by the scrap dealer even prior to placing the cars in the field to await possible parts salvaging. One example of successful salvaging based on this method is the radiator end cap (see Figure 13.5). This part can be cleaned, reground, and molded into useful parts such as drill housings. Similar selective part dismantling may include fan shrouds, fender liners, windshield washer bottles, and interior trim pieces. Of course the person responsible for resin selection for a new application should be aware that such regrind could have properties somewhat inferior to virgin resin, and act accordingly.

As selective dismantling gains momentum, four areas must be addressed before an economical system for the recovery and recycling of used automotive parts can be assured. As defined by the APC, there are four stages to the recycling infrastructure:

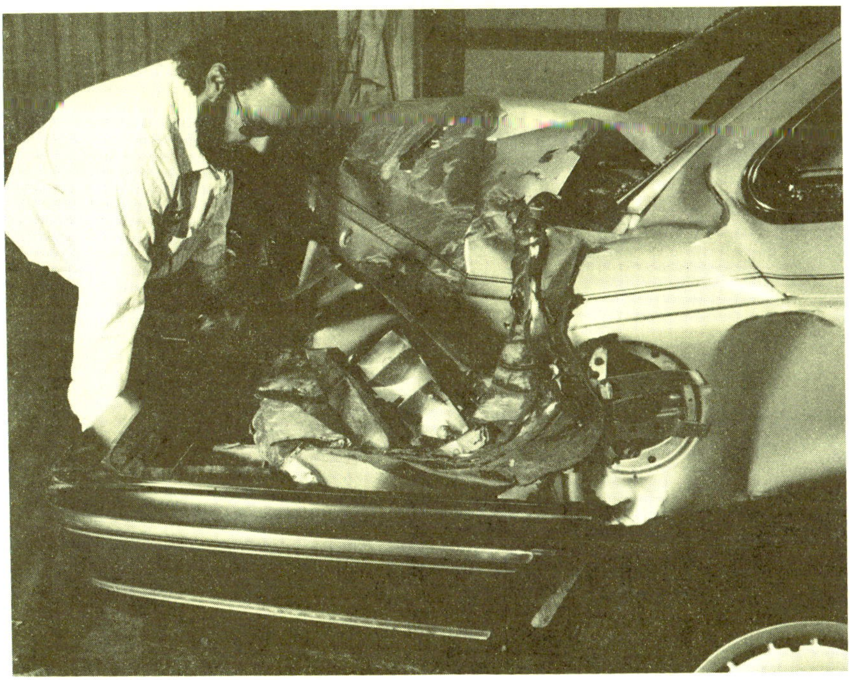

FIGURE 13.6 A plastic bumper is recovered for re-use as a replacement automobile part. (Courtesy of American Plastics Council, Washington, D. C.)

Collection
Handling and sorting
Reclamation
End markets

As mentioned, progress is being made to solve these challenges for the automotive industry, and programs have been established by the APC for the appliance and other industries that use plastics in durable goods.

The selective dismantling operation has been studied during a field demonstration by the APC. "This pilot study examined costs and technical barriers associated with plastics recovery, and is a first step in an ambitious program to substantially improve tomorrow's post automotive recovery level," as stated by Dr. Michael Fisher, director of technology for the APC's Durables Committee. The field demonstration project highlighted several limitations of the dismantling-for-recovery process. It noted that there was a complete lack of design for plastics recycling for every part studied. It further pointed out that very few plastic reclaimers have the capability to process dismantled plastics. However, on the positive side, the demonstration resulted in a technical cost model for the recycling of plastic from scrapped cars, and provides guidance and direction for more in-depth programs in the future.

CODING AND IDENTIFICATION OF PLASTICS

Another, and more progressive, approach to reducing the amount of engineering plastics being landfilled from junked cars is the coding and identification of the plastics used in the parts before the parts are installed in the car. Currently, German automotive companies are leading in this work. They have stated, in effect: "We will not buy your engineering plastics unless you can demonstrate to us a total life cycle plan for your product." There is also an implied threat in the United States, as well as in Japan, that the automotive producers may begin designing parts with aluminum in place of plastic parts, since aluminum is more easily recycled. To forestall these actions, both plastic and automotive producers are taking steps to code and identify plastics so that they can be isolated for recycling easily during the dismantling step.

Since 1991, some automotive manufacturers have begun to select materials for later use in appliances. Manufacturers should be aware that the marking is, in effect, now a consideration of material selection. To become familiar with this aspect of part design they should obtain copies of the standards now in use or about to be used. Standards have been prepared by the American Society for Testing Materials (ASTM), the Society of Automotive Engineers (SAE), and the International Standards Organization (ISO). Since Europe and Japan are leaders in coping with the disposal of plastics from scrapped vehicles, it is quite possible that in the near future the ISO standard will become the standard in the United States as well as abroad.

ECONOMICS OF RECYCLED ENGINEERING PLASTICS

Of course, environmental concerns are the driving force for recycling of durable plastics, but if the process for recycling is not economical it will not happen. Various methods for recovery of economic value from plastics waste are available.[110,111] Major material suppliers are closing in on practical recycling methods. Du Pont, for example, said in 1991 that recycling will definitely change the way it does business, and this will no doubt be true of other major companies. This implies the sale of recycled resins, or possibly a blend of virgin and recycled resins. For these reasons, through the APC, efforts are being made by industry to develop a number of practical and economical applications for recycled resins.

Although the major resin suppliers are becoming more and more involved in returning recycled material to the field, it is unlikely that they will do their own regrinding of the scrapped plastics from durables before processing. Instead, a new industry is rapidly developing to do the regrinding, which calls for designing new machines specifically for the purpose. This would include cutters, extruders, granulators, grinders, conveying systems, and stuffers.

Once the regrind is turned over to the material supplier, it must be processed economically. If mixed resins or contamination are present in the regrind, industry must develop new ways to separate them, probably on a continuous basis. Approaches such as density-gradient or selective melting techniques might be tried. Another method might be to sort out the desired resin from the regrind continuously, using some kind of spectrophotometric or laser-actuated selector. In any case, the need for industry to regrind recycled material is here, and industry will rise to the challenge. At the same time, the one choosing a plastic for a new use must consider the use of recycled resin as well as virgin material.

SUMMARY AND CONCLUSION

It is only within the last few years that governments and industry have realized and faced up to the gravity of the waste disposal problems confronting us today. Landfills by themselves do not constitute a viable means for solving the world's need for a way to dispose of its refuse. It is obvious that appropriate space for landfilling is becoming harder to find, and the waste disposal operation itself is becoming more expensive and more harmful to the environment. Further, newly designed equipment is rapidly being made available that contributes to the economical recovery of reworked polymers. Progress has been made in this area, but of course much more needs to be done. However, it is only fairly recently that intensive effort has been given to the reduction of the volume of durable plastics that has been going into landfills.

The automotive industry is a leader in confronting this problem. With pressure being applied from the governments and from the industry itself, research to find ways of reducing durables in landfills is underway. Some of the approaches to this include selective dismantling of junked cars, separating plastics from shredded cars, and reducing the kinds of plastics in new car designs, thereby making it easier to carry out selecting dismantling.

Designers and end users who select thermoplastics for engineering applications must at all times keep in mind this critical need for an intelligent approach to the reduction of the volume of durable plastics to landfills. In the long run, it is possible that only a few kinds of polymers will be required for automotive and other isolated engineering uses, but that will only happen when joint efforts of designers and resin producers have been able to accomplish it. Until then the selection process must be carefully thought out.

14

Other Engineering Thermoplastics

INTRODUCTION

In earlier chapters the point was made that it is impractical to cover individually every engineering thermoplastic composition available to the selector; for this reason, only 15 of higher volume were discussed in any detail. It was also mentioned that, in addition to the common high-volume base polymers, a number of newer polymers have attractive properties to consider. For details regarding properties and engineering aspects of various plastics, readers can refer to some of the available handbooks on the subject.[118–123]

In general, material suppliers have concentrated on new polymers that, even in the unreinforced state, have higher melting points, greater stiffness, and higher heat-deflection properties than the high-volume resins discussed. In addition, most of the new polymers exhibit some unique properties, such as unusually high end-use temperatures over long periods of time, or unusual resistance to the effects of chemicals and solvents. Further, properties of most of them respond to improvement in stiffness and creep properties due to incorporation of glass fiber reinforcement.

In general, they also share the distinction of having higher prices than the high-volume polymers, which is natural enough. Perhaps one other difference is that most of the new polymers have high melting points, which, in end-use situations, is normally very desirable. On the other hand, although by no means a serious problem, the higher melting points tend to discourage some of the less sophisticated injection molders.

No attempt to rank these newer polymer in any way has been made within their own grouping or with the larger-volume resins. Instead, the properties are listed and a short description of each is included.

Perhaps because of the newness of these polymers, some of the physical property data are not available in the literature, and it was necessary at times to estimate or extrapolate data obtained from other sources. This is particularly true for creep modulus, which simply is not uniformly reported by the various material suppliers, so the creep data shown must not be used for any design calculations. The values as shown serve only to indicate a range.

A summary of the physical property data for the resins is given in Table 14.1.

TEFLON®* PFA FLUOROCARBON RESIN

Chemistry

Du Pont's PFA fluorocarbon resin is a copolymer of tetrafluorethylene and perfluoro vinyl ethers (see Figure 14.1).

PFA was not included in the preceding ranking exercise for engineering thermoplastics because it has low tensiles, low modulus, and high creep properties compared to the 15 members of "the family." Nevertheless, it has high and low end-use temperature properties for difficult technical applications requiring injection-molded parts, and is definitely a member of the new and specialized group of engineering thermoplastics.

Properties

PFA is used in applications where very few other materials are suitable. In this context, its important properties are:

Continuous use temperatures of 500°F
Unaffected by weather

*Teflon® is a trademark of E. I. du Pont de Nemours & Co., Inc., Wilmington, DE.

TABLE 14.1 Physical Property Data for the Resins

Resin	Tensile strength (kpsi) ASTM D-638	Flexural modulus (kpsi) ASTM D-790	Heat-deflection temperature (°F @ 264 psi) ASTM D-648	Notched Izod (ft·lb/in.^{2}) ASTM D-256
Fluorocarbon copolymer (Teflon® PFA)	4.0	95	124	No break
Polyether sulfone	12.2	373	400	1.6
Polyarylate	10.0	310	345	4.2
Polyphenylene sulfide (40% glass)	17.5	1800	500	1.4
Polyamide-imide	26.9	664	525	2.5
Polyether-imide	15.2	480	392	1.0
Resin	**Elongation at break (%) ASTM D-638**	**Creep modulus (kpsi)**	**Dielectric strength (V/mil) ASTM D-149**	**Arc resistance (s) ASTM D-495**
Fluorocarbon copolymer (Teflon® PFA)	200–600	34	2000	240
Polyether sulfone	40–80	NA[a]	400	150
Polyarylate	50	200	400	125
Polyphenylene sulfide (40% glass)	1	1200	450	35
Polyamide-imide	12–18	400	600	4
Polyether-imide	60	400	480	126
Resin	**Specific gravity ASTM D-792**	**Approximate melting point (°F)**	**Approximate average shrinkage (in.)**	**Cost[b] (¢/in.^{3})**
Fluorocarbon copolymer (Teflon® PFA)	2.15	580	10–20	148
Polyether sulfone	1.37	650	6	29
Polyarylate	1.21	650	9	11
Polyphenylene sulfide (40% glass)	1.68	545	2–3	17
Polyamide-imide	1.40	527	6–8	128
Polyether-imide	1.27	527	5–7	22

[a]Not available.
[b]1996.
Source: Ref. 105.

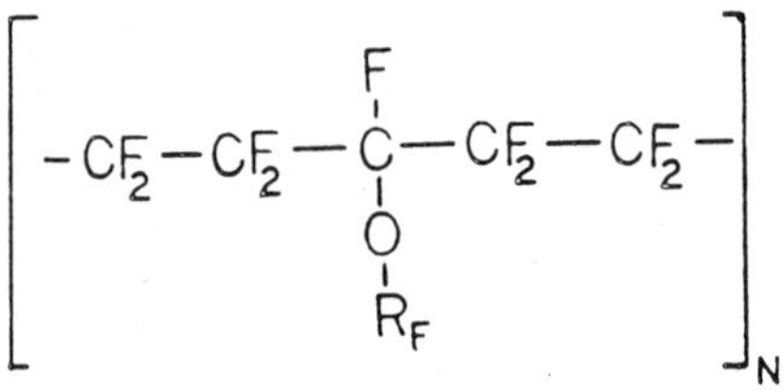

FIGURE 14.1 Chemical structure of the fluorocarbon copolymer Teflon® PFA.

Excellent electricals
Very low coefficient of friction
Useful at cryogenic temperatures
Outstanding resistance to chemicals

End Uses

As might be expected, PFA is a unique material. It is also expensive and not a high-volume resin compared to the more commonly used engineering plastics such as acetals, polycarbonates, nylons, etc.

Because of its chemical inertness, PFA is used in molded articles exposed to very corrosive chemicals. Examples are molded laboratory ware, seals, valve liners, pump fittings, and pipe fittings.

Since PFA retains its electrical properties over a broad temperature range, it is used for insulators or sophisticated electronic equipment such as remote radar installations. It has also found use in stand-off insulators, as well as spacers in large-diameter electrical conduits.

Special Considerations

To obtain the best properties from PFA moldings, it is necessary to minimize or, preferably, eliminate voids in the part because the effect of even microvoids can result in deleterious effects on fatigue life and gas permeability. PFA has high creep, and parts must be carefully designed to take this into consideration.

Since PFA has nearly 10 times the thermal expansion of steel, care must be exercised during part design, if such a differential is significant.

POLYETHER SULFONE

Chemistry

Unlike polysulfone, which contains an alkyl component, polyether sulfone is a wholly aromatic ether sulfone (see Figure 14.2).

Polyether sulfone is an amorphous polymer. It is marketed by ICI Americas, Inc., under the trademark Victrex®.

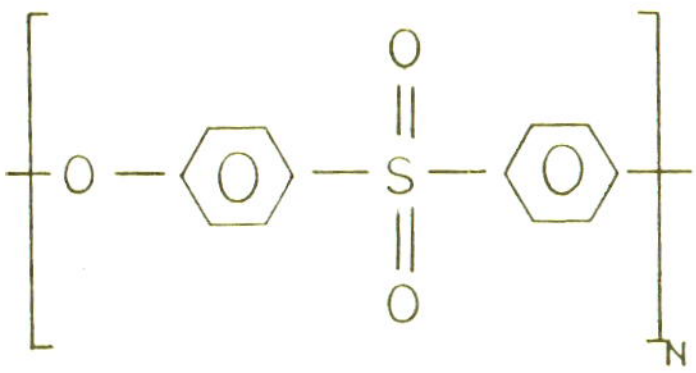

FIGURE 14.2 Chemical structure of polyether sulfone.

Properties

Like a number of this group of engineering plastics, polyether sulfone has a number of good properties, with the emphasis on high-temperature applications.[29] It has

- Outstanding thermal stability
- High heat-deflection temperature
- Very good electricals
- Good dimensional stability over a range of temperatures

End Uses

End uses include:

- High-temperature electrical multipin connectors
- Printed-circuit boards
- Bearing cages
- High-temperature personal product appliances (e.g., hair dryer outlets, hot combs, projector grill jambs)
- Some medical applications requiring sterilization

Special Considerations

At 29 cents/in.^{3}, polyether sulfone is competitive in price to polyether-imide and glass-reinforced polyphenylene sulfide.

A number of solvents will adversely affect this amorphous polymer, particularly if the molded part is under stress. It is not very resistant to the effects of ultraviolet light.

POLYARYLATE

Chemistry

Polyarylates are aromatic polyesters of phthalic acids and bis-phenols, and are amorphous in structure (see Figure 14.3, which shows the structure of two components of a typical arylate copolymer).

The major manufacturer of polyarylates is Union Carbide, under the trademark Ardel®.

Properties

Ardel® D-100 is one of the first commercial compositions in the new family of polyarylates, and some commercial literature is available on it. It also has some interesting properties:

Good toughness
Very good weatherability
Good flexural recovery
High heat-deflection temperature
Good electricals
VO without additives
Good creep properties

End Uses

End uses include:

Electronic and electrical hardware
Solar energy collector components
Lighting applications
Safety devices requiring UV resistance and transparency with toughness
Transportation industry

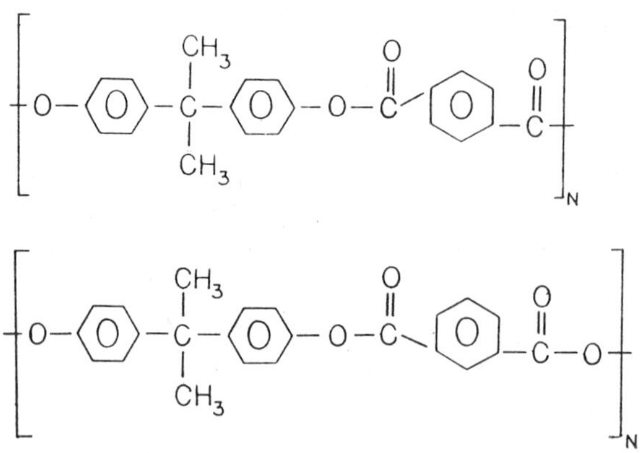

FIGURE 14.3 Chemical structure of Ardel® polyarylate D-100.

Special Considerations

Polyarylates are not particularly resistant to stress cracking in the presence of a number of organic solvents, and properties of the resins deteriorate with immersion in water over a fairly short period of time.

At 11 cents/in.^{3}, it is the least expensive of these newer engineering thermoplastics.

POLYPHENYLENE SULFIDE

Chemistry

The structure of polyphenylene sulfide is simple, as illustrated in Figure 14.4. It is sold only in the 40% glass-reinforced state by Phillips Chemical Company under the trademark Ryton®.

This aromatic polymer is crystalline in form.

Properties

Polyphenylene sulfide in the unreinforced state lacks some desirable properties, particularly sufficient toughness. However, as sold in the glass-reinforced state, the toughness is adequate.[33] Other properties are:

Outstanding chemical resistance, second only to polytetrafluoroethylene
Very high modulus
Outstanding creep resistance
Good electricals
Excellent heat-deflection temperature

End Uses

With its chemical resistance and high-temperature properties, the end uses of polyphenylene sulfide are predictable:

Mechanical applications in chemical processing systems
Electrical/electronic uses such as computer connectors, coil forms, and bobbins
High-voltage applications

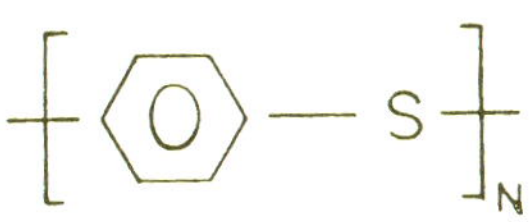

FIGURE 14.4 Chemical structure of polyphenylene sulfide.

Special Considerations

As is the case with most of these polymers, the price per cubic inch (at 17 cents/in.[3]) is significantly higher than the average for the 15 more common resins discussed.

In processing, a good surface can be obtained by maintaining a mold temperature of 250–275°F, but if optimum toughness is required, a temperature for the mold of 150°F is preferable.

POLYAMIDE-IMIDE

Chemistry

Polyamide-imide is a condensation polymer of trimellitic anhydride and various aromatic diamines, with the general chemical structure shown in Figure 14.5.

It is an injection moldable, high-performance engineering thermoplastic marketed by Amoco Chemicals Corporation under the trade name Torlon®.

Properties

The polyamide-imide polymer has a number of attractive properties:

Excellent tensile strength
High stiffness
Useful properties at cryogenic temperatures
Excellent creep resistance
Meets VO flame requirements
Good solvent and chemical resistance
High heat-deflection temperature

End Uses

Because of its high heat-deflection temperature and its good electricals, polyamide-imide is used in some demanding applications, such as:

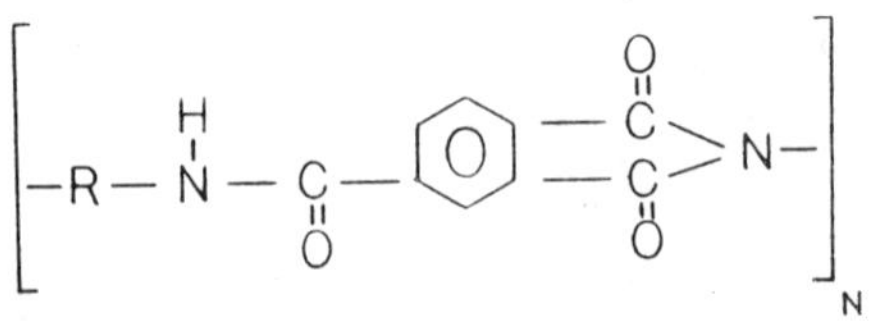

FIGURE 14.5 Chemical structure of polyamide-imide.

Jet engine components
Compressor and generator parts
Electronic/electrical applications where resistance to high heats are needed
Hydraulic/pneumatic industries for seals and bearings

Special Considerations

Polyamide-imide is unaffected by aliphatic and aromatic hydrocarbons, halogenated solvents, and most acids and bases at normal ambient temperatures. At elevated temperatures, it is affected by steam, caustic, and some acids.

As might be expected, it absorbs some moisture at normal 50% RH, 70°F conditions, going to 1% in 100 hours in such an environment.

Although this polymer offers some attractive properties, it is expensive at 128 cents/in.^{3} as compared, for example, to the glass-reinforced polyesters at an average price of about 8.7 cents/in.^{3}

POLYETHER-IMIDE

Chemistry

Polyether-imide is a complex polymer first marketed in 1982 by the General Electric Company under the trademark Ultem® (see Figure 14.6).

Polyether-imide is an amorphous polymer.

Properties

Polyether-imide is a very useful thermoplastic, with a good blend of properties:

High tensile strength and stiffness
Excellent electrical properties
Good creep resistance
High heat-deflection temperature
Good thermal stability and warp resistance at elevated temperatures

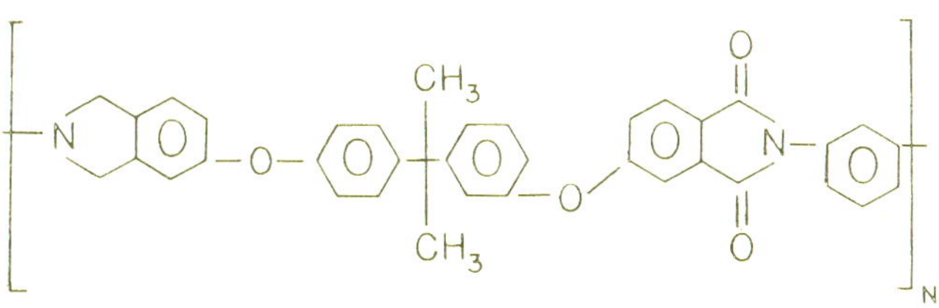

Figure 14.6 Chemical structure of polyether-imide.

End Uses

End uses include:

Electrical/electronic parts, such as high-voltage current breaker housings
Jet engine components
Aircraft interior seating components
Under-the-hood automotive applications

Special Considerations

Polyether-imide has good resistance to most organic solvents and dilute acids and bases. Although it is generally resistant to completely halogenated compounds, it is affected by partially halogenated products such as trichloroethylene.

The product is somewhat notch sensitive, with a notched Izod value of about 1.0. At 22 cents/in.^{3}, it is more expensive than any of the "family" of 15 common engineering thermoplastics.

15

Computerized Approach to Resin Selection

INTRODUCTION

With the growing importance of the use of computer assistance in every aspect of design and fabrication of plastics products, it is natural to expect efforts directed toward the use of computers in resin selection.

The range of physical properties of available thermoplastics is so broad that even the most experienced materials engineer has at best only a general idea of the structural potential of whole families of materials. A number of computer-assisted materials selection (CAMS) databases have been created with the idea of providing quick access to current, comprehensive, and accurate characteristics of the widest choice of candidate materials.[124]

An example of such a system is Plastic Technology's Plaspec, which is programmed to provide a list of resins filling the mechanical, electrical/optical, processing, or cost requirements typed in by the user. Similarly, Plastiserve is also a database of several thousand resins, made available through the Compuserve network and developed jointly with Borg-Warner Chemicals.

As the computer age continues to mature, the concept of applying computer programs to select a specific resin for an engineering application would seem to be logical. A number of very capable material supplier computer programmers have tackled the problem, but after several years have shelved or discontinued their efforts entirely.

There is speculation as to why the effort was abandoned. Perhaps one reason is the parochial nature of the problem; that is, if a truly efficient program developed by one supplier led a resin selector to choose a resin supplied by a competitor, it certainly would be nonproductive for the first supplier.

Another reason appears to involve the need for the means to quantify the combination of properties of a resin that would be meaningful. For example, a multiple regression analysis approach would seem to be reasonable. Numerical values for such independent variables as tensile strength, toughness, stiffness, etc., are simple to obtain, and could be applied to a multiple correlation. But what value should be ascribed to the dependent variable that could be interpreted as the optimum resin selection? If this kind of an approach has been considered, it appears that success has been elusive.

Perhaps another reason computerized material selection was abandoned is the zeal applied to attaining perfection in the final choice. Obviously, many of the important physical property characteristics of plastics vary considerably with time, temperature, load, etc., and to quantify the variables accurately for very specific conditions would require a vast amount of data, considering how many possible polymers and blends might be under review.

Actually, mathematical perfection in such a search is not really a practical objective. Indeed, if the goal had been to narrow the selection down to just a few resins and then use experience and empirical observation to finalize the best choice, perhaps success would have been obtained.

The practical difficulty is, of course, that experienced material selectors and experts are often not readily available. In addition, there are complexities in identifying the most significant end-use requirements and comparing the characteristics of the narrowed list of resins in order to make the best final choice. A casual approach to resin selection even at this stage may lead to the wrong choice from the narrowed list. Trial-and-error alternatives can be very expensive.

A more convenient and reliable alternative to the above-mentioned screening matrix approach is a "knowledge-based expert system", built through the application of artificial intelligence principles. In very simple terms, an expert system is a computer program which captures the experience, knowledge, and judgment of expert practioners in an organized manner, and makes that information available to inexperienced personnel. A limited number of applications of expert systems in polymer science and engineering have been explored.[52,53,125,126]

GENERAL CONCEPTS OF AN EXPERT SYSTEM

An expert system is a highly proficient computer-based problem solver which embodies specialized knowledge in a narrow, specific area of practical importance. It relies entirely on declarative and procedural knowledge for data processing. Declarative knowledge is information which is known through facts, theories, rules, hypotheses, or heuristics. On the other hand, procedural knowledge provides instructions in terms of strategies and tactics on how to use the declarative knowledge.

Expert systems have found a wide variety of applications, including selection, interpretation, prediction, diagnosis, debugging, repair, design, planning, monitoring, control, and instruction. Typical examples of each type of application do exist, as well as those which perform the tasks of two or three of the activities together. Medical science received the most attention by expert system developers. The functional capabilities of the expert system MYCIN are well known.[127,128] Several other expert systems in other disciplines such as chemistry and engineering are in commercial use and are known to perform at levels comparable to or even better than human experts.

A number of books have been written about expert systems that discuss most of the concepts in vivid detail. In the following text, a brief descriptions of some important aspects are presented. Those wishing to learn more about expert systems should refer to specialized books on the subject.[129–139]

Architecture of an Expert System

The architecture of an expert system can be well understood by broadly visualizing the system as consisting of three major parts: the knowledge base, the inference engine, and the user interface.

Knowledge Base

The knowledge base, which holds the knowledge about the domain, is developed through the use of one or more of the standard representation techniques, namely, production rules, semantic networks, frames, or formal logic. In principle, any of the techniques can be used for knowledge representation. However, the case of manipulating and accessing the knowledge varies from problem to problem depending on its intrinsic nature. In the case of material selection, the knowledge representation is done most conveniently through the use of production rules. Hence, only this type of representation is discussed further.

Production rules are built by using IF/THEN type statements. These clauses are meant to mimic the way experts store their domain knowledge following "rules of thumb" or heuristics with cause–effect structure. Each rule is made up of one or more conditions which, IF satisfied, THEN forces one or more actions. The actions of a rule may add new facts, remove old ones, or modify existing ones in the knowledge base. The modified knowledge is then used for matching the IF conditions in the other rules, which in turn leads to other actions. This process continues until the system can find no rules with IF clauses that can be matched successfully. At this stage, the expert system presents the results.

Step-by-step procedures for developing expert systems are available[140–143] and should be referred to before undertaking the task. In order to build the production rules, any of the commercially available expert system development environments can be used. Selection criteria for expert system shells is available[144] and should be referred to for details.

Inference Engine

The inference engine is the reasoning capability of the expert system. It has the ability to select the appropriate portions of the knowledge base depending on the problem at hand and then reach a conclusion through deductive logic. Its reasoning procedure is based on either a backward or a forward chaining technique. Backward chaining is a goal-directed technique, while forward-chaining is a data-driven technique. The inference engine often has the ability to deal with problems where the data are uncertain, and in such circumstances can provide conclusions expressed in degrees of certainties or probabilities.

User Interface

The user interface is the part of the expert system that interacts with the end user. It acts by asking the right questions, presenting the results, and providing explanations as to why and how particular actions are taken. The explanation facility gives the end user an insight into the expert system's reasoning process and thereby allows the option to accept or reject a particular recommendation.

EXPERT SYSTEMS IN MATERIALS SELECTION

Polymer Selection Assistant (PSA) was among the first expert system which was attempted,[125] using a combination of dBase II databases along with the knowledge base of the shell INSIGHT 2. The knowledge base invokes the expert system that asks a series of questions to determine the possible engineering use requirements for the plastic product. The expert system then activates several DBPAS Pascal programs that fetched data concerning the characteristics of 13 different plastics that were stored in the dBase II database POLYMER.DBF.

The main goal is to find if the polymer is appropriate based on the aptness of each characteristic such as engineering use temperature, hardness, density, thermal conductivity, flammability, acid resistance, base resistance, solvent resistance, tensile strength, compression strength, flexural strength, elongation, clarity, and impact strength. The expert system calculates a rating for each of the plastic's aptness in relation to the desired end use. This calculated rating is saved in the POLYMER.DBF database. After the expert system has evaluated all the plastics in the database, it invokes another DBPAS Pascal program that prints a report of its findings.

Another expert system was suggested for assisting plastic parts designers in designing moldable parts as well as picking the appropriate resin for the job.[126] It was called MAPS, for Material And Process Selection. Since it was only an experimental program, it contained a limited database of a small number of resins with information on specific material properties of each.

MAPS has been programmed to interact with the user/designer, asking questions that will cause key requirements to surface, and focusing the dialogue along the lines of information supplied by the designer. For example, if the designer knows the required heat-deflection temperature MAPS will not ask further questions concerning HDT. However, if the designer does not know the required HDT, MAPS will attempt to answer the question itself by asking the designer more basic questions about the environment the part will encounter. MAPS basically tries to determine the most restrictive requirements of the part being designed, thereby reducing the number of possible solutions and thus the number of questions that must be asked.

MAPS first asks questions about the environment and mechanical requirements of the part and then searches the resin database for resins with properties matching those in the requirements. After MAPS completes its search, one of the three cases exists—no resin, a large number of resins, or a small number of resins meet the requirements. In the first two cases, MAPS suggests a rerun of the program with relaxed or restricted conditions, respectively. Only when the list of possible candidate resins is reasonably small is the selection assumed to be complete.

PSA and MAPS both operate on a credit-assignment approach. But the suggested list of resins is not given in prioritized order based on the ranking of their characteristics. The method of ranking of the resins in the manner shown in Chapter 11 is very useful to determine the most appropriate resin from a narrowed list. In order to demonstrate this, a new expert system sample program named SELECTHER has been developed for SELECTing THERmoplastics based on their property rankings. In the remainder of this chapter, a step-by-step procedure is given for building, testing, and using the expert system SELECTHER.

GENERAL DESCRIPTION OF SELECTHER

SELECTHER is a forward-chaining production system that embodies diagnostic knowledge in the form of a modular collection of IF/THEN rules. When a certain situation exists (i.e., when the state in the IF portion is satisfied on account of the inputs provided by the user to the requested information), the production rule is executed (i.e., the action in the THEN portion is taken). The production system evaluates the rules in the correct sequence and recommends the necessary actions after appropriately choosing the right rules to fire from among many candidate rules.

In this section, a complete description of SELECTHER is given, from its knowledge base to its treatment methodology and including certain characteristic features.

Knowledge Base

The knowledge base of SELECTHER is essentially extracted by handcrafting through the compiled information from the preceding chapters, particularly, Chapters 2 and 11. Before writing the production rules, the available knowledge was reorganized to be amenable to the formation of IF/THEN rule sets. The knowledge thus reorganized is presented below.

The Price of the Resin [P_Resin] must be less than the Estimated Cost of the Product [ECP]. All resins costing more than [ECP] have to be eliminated from consideration during the selection process.

The MAXimum End-use Temperature [MAX_ET] must be less than the Heat-Deflection Temperature of the Resin [HDT_Resin]. All resins with heat-deflection temperature less than [MAX_ET] have to be eliminated from consideration during the selection process.

In case the engineering thermoplastic product during end-use application is likely to be exposed to any particular chemicals/solvents, then all resins which are affected by the particular chemicals/solvents have to be eliminated from consideration during the selection process. Those resins which are slightly affected may be carried forward for further consideration but with a confidence factor of only 50%. Those resins which are not affected at all are considered with a confidence factor of 100%.

If the product is held frequently at elevated temperatures for extended periods of time, then caution is to be exercised because heat aging can cause eventual deterioration of polymer properties or even complete failure of its structure.

If the product is held frequently at temperatures below freezing (less than 32°F), then caution is to be exercised because impact resistance and toughness are reduced drastically even by a drop of a few degrees in temperature in this range.

If the engineering thermoplastic product during end-use application is at all times immersed in water whose temperature exceeds 125°F, then PET and PC have to be eliminated from further consideration because they lose properties significantly when exposed to hot water for long periods of time.

If the ambient relative humidity (where the product is being manufactured and/or being used) is greater than 50% and changes frequently with time, then nylons have to be eliminated from consideration because there will be dimensional changes in the molded part besides deterioration of structural properties such as stiffness and strength.

If the proposed application involves the use of metals and the thermoplastic as part of a single item, then care has to be taken to allow for differences in the coefficient of thermal expansion values between the two, when there is likelihood of temperature changes.

If good product appearance and gloss is of critical importance, then resin selection has to be limited to the natural resins because a heavily loaded glass-reinforced resin is not going to have the gloss of a resin without glass.

Based on the structural requirements, the tensile strength, flexural modulus, notched Izod, elongation at break, and creep modulus must have minimum values or need to be specified in a particular range. For example, tensile strength of less than 5 kpsi is normally not considered for most engineering plastic applications. Therefore, ABS and PP (with 40% talc) would have to be eliminated on the basis of this property (see resin ranking based on tensile strength values in Chapter 11). Similarly, only an Izod over 1.5 represents a resin that is reasonably tough. This would mean the elimination of acetals, mineral-reinforced nylon, and talc-filled PP from consideration on the basis of this property (see resin ranking based on notched Izod values in Chapter 11).

Similarly, based on the electrical requirements, the dielectric strength and arc resistance need to be specified in a particular range.

Rules Structure with Probability Assignments

With the knowledge base in the reorganized form as given above, a strategy had to be planned for structuring the production rules. For the creation of SELECTHER, the expert system shell[144] used is EXSYS which comes with a development editor and debugging aid called EDITXS.* Since the production rules are built in the IF/THEN format, the various requirements that the resins must satisfy are included in the IF clauses and the resins which satisfy the requirements are inserted in the THEN action clauses. Each resin is accompanied by a PROBABILITY value based on the certainty of the resin's ability to satisfy all the various requirements in different degrees.

When assigning the probability values, it has been assumed that the total list of resins considered in the selection process is exhaustive. SELECTHER is built using a total of 15 ranked resins. Hence the difference in the probability value between each sequentially ranked resin is taken as 100/15 ≈ 7. Thus, the resin with

Rank 1 is given the Probability	99%
Rank 2 is given the Probability	92%
Rank 3 is given the Probability	85%
Rank 4 is given the Probability	78%
Rank 5 is given the Probability	71%
Rank 6 is given the Probability	64%
Rank 7 is given the Probability	57%
Rank 8 is given the Probability	50%
Rank 9 is given the Probability	43%
Rank10 is given the Probability	36%
Rank11 is given the Probability	29%
Rank12 is given the Probability	22%
Rank13 is given the Probability	15%
Rank14 is given the Probability	8%
Rank15 is given the Probability	1%

Every resin is given a different probability value for each property because it has a different rank for a different property.

When a resin satisfies all requirements, it is included in the chosen list and assigned a probability which is the average of the probabilities of all the properties. The process is similar to what was done in Table 11.17 in order to determine the general usefulness ranking.

The rules structure and the probability assignments are best understood by actually perusing some of the production rules of SELECTHER.

Some Sample Production Rules

The actual method of building the production rules in EXSYS through its editor and debugging aid EDITXS should be studied from the manual of the

*Development environment and editor available from EXSYS, Inc., P.O. Box 75158, Contract Station 14, Albuquerque, NM 87194, USA.

expert system development environment. However, a list of qualifiers, choices, and variables are given in Appendix B. Similarly, the contents of various files that need to be prepared for the working version of the expert system SELECTHER are given in Appendix B.

In the following, some of the production rules are given so that it will be easy for interested readers to build their own expert system for a new set of resins or for a longer list of resins.

```
Subject:
                  S E L E C T H E R
    Expert System Demonstration Program for Selecting
    Thermoplastics for Engineering Applications

Uses all applicable rules in data derivations.

RULES:
_______________________________

RULE NUMBER: 1

IF:
     [P_ABS] < [ECP]
 and [MAX_ET] < [HDT_ABS]
 and The Engineering Thermoplastic product during end-use
     application is likely to be exposed to the following
     chemicals/solvents: NOT Aromatics
 and The Engineering Thermoplastic product during end-use
     application is held at all times at normal temperatures of
     about 73 degrees Fahrenheit or held frequently at elevated
     temperatures for extended period of time or held frequently
     at temperatures below freezing (less than 32 degrees
     Fahrenheit) for extended period of time
 and The Engineering Thermoplastic product during end-use
     application is at all times immersed in water whose
     temperature exceeds 125 degrees Fahrenheit or immersed in
     water whose temperature is normally at around 73 degrees
     Fahrenheit or not immersed in water
 and The ambient relative humidity (where the product is being
     manufactured and/or being used) is greater than 50% and
     changes frequently with time or greater than 50% and does
     not change with time or less than 50% and changes frequently
     with time or less than 50% and does not change with time
 and The proposed application involves the use of metals and the
     thermoplastic as part of a single item: No
 and The Tensile Strength (TS) value in K p.s.i. needs to be in
     the range of 4 <= (TS) < 23
 and The Flexural Modulus (FM) value in K p.s.i. needs to be in
     the range of 100 <= (FM) < 1400 or 300 <= (FM) < 1400
 and The Notched Izod (NI) value at 73 degrees F in ft-lb/in
     needs to be in the range of 0.4 <= (NI) < 50 or 1.5 <= (NI)
     < 50 or 3 <= (NI) < 50
```

and The Elongation at Break (EB) value in percentage needs to be in the range of 3 <= (EB) < 400 or 5 <= (EB) < 400 or 7 <= (EB) < 400 or 10 <= (EB) < 400 or 25 <= (EB) < 400

and The Creep Modulus (CM) value in K p.s.i. needs to be in the range of 35 <= (CM) < 1400 or 50 <= (CM) < 1400 or 75 <= (CM) < 1400 or 150 <= (CM) < 1400

and The Dielectric Strength (DS) value in V/mil needs to be in the range of 350 <= (DS) < 600 or 400 <= (DS) < 600

and The Arc Resistance (AR) value in sec. needs to be in the range of 5 <= (AR) < 250 or 50 <= (AR) < 250 or 75 <= (AR) < 250

THEN:

[ABS] IS GIVEN THE VALUE 1

NOTE:
This rule will evaluate whether ABS is an OKAY choice based on its price, maximum end-use temperature, anticipated environment and structural requirements in use.

REFERENCE:

Charles P. MacDermott and Aroon V. Shenoy, Selecting Thermoplastics for Engineering Applications, Marcel Dekker Inc., New York, 1997 (Tables in Chapter 11)

RULE NUMBER: 2

IF:

[ABS] = 1

and [P_ABS] < [ECP]

THEN:

ABS - Probability = 92/100

NOTE:
P_ABS = 3.4 is the Price of ABS in cents per cubic inch and ECP is the Expected Cost of Product in cents per cubic inch. It is ranked no. 2 for Price.

REFERENCE:
Charles P. MacDermott and Aroon v. Shenoy, Selecting Thermoplastics for Engineering Applications, Marcel Dekker Inc., New York, 1997 (Table 11.14)

RULE NUMBER: 3

IF:

[ABS] = 1

and [MAX_ET] < [HDT_ABS]

THEN:
    ABS - Probability = 29/100

NOTE:
MAX_ET is the MAXimum End-use Temperature in degrees Fahrenheit and HDT_ABS = 216 is the Heat Deflection Temperature of ABS at 264 p.s.i. in degrees Fahrenheit. It is ranked no. 11 for HDT.

REFERENCE:
Charles P. MacDermott and Aroon V. Shenoy, Selecting Thermoplastics for Engineering Applications, Marcel Dekker Inc., New York, 1997 (Table 11.6)

RULE NUMBER: 4

IF:
    [ABS] = 1
  and The Engineering Thermoplastic product during end-use application is likely to be exposed to the following chemicals/solvents: Strong acid or Strong base or Salt solution or None of the above or Unknown

THEN:
    ABS - Probability = 100/100

NOTE:
ABS is not affected by Strong acid or Strong base or Salt solution at room temperature.

REFERENCE:
Charles P. MacDermott and Aroon V. Shenoy, Selecting Thermoplastics for Engineering Applications, Marcel Dekker Inc., New York, 1997 (Table 11.2)

RULE NUMBER: 5

IF:
    [ABS] = 1
  and The Engineering Thermoplastic product during end-use application is likely to be exposed to the following chemicals/solvents: Gas or Alcohol or Ketone or Aldehyde or None of the above or Unknown

THEN:
    ABS - Probability = 50/100

NOTE:
ABS is slightly affected by Gas or Alcohol or Ketone or Aldehyde at room temperature.

REFERENCE:
Charles P. MacDermott and Aroon V. Shenoy, Selecting Thermoplastics

for Engineering Applications, Marcel Dekker Inc., New York, 1997 (Table 11.2)

RULE NUMBER: 6

IF:

[ABS] = 1

and The Tensile Strength (TS) value in K p.s.i. needs to be in the range of 4 <= (TS) < 23

THEN:

ABS - Probability = 8/100

NOTE:
Tensile Strength is to be measured under ASTM D-638. (TS) value of less than 5 K p.s.i. is not normally considered for most engineering thermoplastic applications. (TS) value for ABS = 5 K p.s.i. It is ranked no. 14 for this property.

REFERENCE:
Charles P. MacDermott and Aroon V. Shenoy, Selecting Thermoplastics for Engineering Applications, Marcel Dekker Inc., New York, 1997 (Table 11.3)

RULE NUMBER: 7

IF:

[ABS] = 1

and The Flexural Modulus (FM) value in K p.s.i. needs to be in the range of 100 <= (FM) < 1400 or 300 <= (FM) < 1400

THEN:

ABS - Probability = 36/100

NOTE:
Flexural Modulus is to be measured under ASTM D-790. (FM) value for ABS = 370 K p.s.i. It is ranked no. 10 for this property.

REFERENCE:
Charles P. MacDermott and Aroon V. Shenoy, Selecting Thermoplastics for Engineering Applications, Marcel Dekker Inc., New York, 1997 (Table 11.4)

RULE NUMBER: 8

IF:

[ABS] = 1

and The Notched Izod (NI) value at 73 degrees F in ft-lb/in needs to be in the range of 0.4 <= (NI) < 50 or 1.5 <= (NI) < 50 or 3 <= (NI) < 50

THEN:

ABS - Probability = 85/100

NOTE:
Notched Izod at 73 degrees F is to be measured under ASTM D-256. An Izod over 1.5 represents a resin that is reasonably tough. (NI) value for ABS = 6 ft-lb/in. It is ranked no. 3 for this property.

REFERENCE:
Charles P. MacDermott and Aroon V. Shenoy, Selecting Thermoplastics for Engineering Applications, Marcel Dekker Inc., New York, 1997 (Table 11.7)

RULE NUMBER: 9

IF:

[ABS] = 1

and The Elongation at Break (EB) value in percentage needs to be in the range of 3 <= (EB) < 400 or 5 <= (EB) < 400 or 7 <= (EB) < 400 or 10 <= (EB) < 400 or 25 <= (EB) < 400

THEN:

ABS - Probability = 50/100

NOTE:
Percentage Elongation at Break is to be measured under ASTM D-638. (EB) value for ABS = 5-25%. It is ranked no. 8 for this property.

REFERENCE:
Charles P. MacDermott and Aroon V. Shenoy, Selecting Thermoplastics for Engineering Applications, Marcel Dekker Inc., New York, 1997 (Table 11.8)

RULE NUMBER: 10

IF:

[ABS] = 1

and The Creep Modulus (CM) value in K p.s.i. needs to be in the range of 35 <= (CM) < 1400 or 50 <= (CM) < 1400 or 75 <= (CM) < 1400 or 150 <= (CM) < 1400

THEN: ABS - Probability = 22/100

NOTE:
(CM) value for ABS = 215 K p.s.i. It is ranked no. 12 for this property.

REFERENCE:
Charles P. MacDermott and Aroon V. Shenoy, Selecting Thermoplastics for Engineering Applications, Marcel Dekker Inc., New York, 1997 (Table 11.9)

RULE NUMBER: 11

IF:

[ABS] = 1

and The Dielectric Strength (DS) value in V/mil needs to be in the range of 350 <= (DS) < 600 or 400 <= (DS) < 600

THEN:

ABS - Probability = 22/100

NOTE:

Dielectric Strength is to be measured under ASTM D-149. (DS) value for ABS = 420 V/mil. It is ranked no. 12 for this property.

REFERENCE:

Charles P. MacDermott and Aroon V. Shenoy, Selecting Thermoplastics for Engineering Applications, Marcel Dekker Inc., New York, 1997 (Table 11.10)

RULE NUMBER: 12

IF:

[ABS] = 1

and The Arc Resistance (AR) value in sec. needs to be in the range of 5 <= (AR) < 250 or 50 <= (AR) < 250 or 75 <= (AR) < 250

THEN:

ABS - Probability = 36/100

NOTE:

Arc Resistance is to be measured under ASTM D-495. (AR) value for ABS = 83 sec. It is ranked no. 10 for this property.

REFERENCE:

Charles P. MacDermott and Aroon V. Shenoy, Selecting Thermoplastics for Engineering Applications, Marcel Dekker Inc., New York, 1997 (Table 11.11)

RULE NUMBER: 13

IF:

[P_ACETAL] < [ECP]

and [MAX_ET] < [HDT_ACETAL]

and The Engineering Thermoplastic product during end-use application is likely to be exposed to the following chemicals/solvents: NOT Strong acid or Strong base

and The Engineering Thermoplastic product during end-use application is held at all times at normal temperatures of about 73 degrees Fahrenheit or held frequently at elevated temperatures for extended period of time or held frequently at temperatures below freezing (less than 32 degrees Fahrenheit) for extended period of time

and The Engineering Thermoplastic product during end-use application is at all times immersed in water whose temperature exceeds 125 degrees Fahrenheit or immersed in water whose temperature is normally at around 73 degrees Fahrenheit or not immersed in water

and The ambient relative humidity (where the product is being manufactured and/or being used) is greater than 50% and changes frequently with time or greater than 50% and does not change with time or less than 50% and changes frequently with time or less than 50% and does not change with time

and The proposed application involves the use of metals and the thermoplastic as part of a single item: No

and The Tensile Strength (TS) value in K p.s.i. needs to be in the range of 4 <= (TS) < 23 or 7 <= (TS) < 23 or 10 <= (TS) < 23

and The Flexural Modulus (FM) value in K p.s.i. needs to be in the range of 100 <= (FM) < 1400 or 300 <= (FM) < 1400 or 400 <= (FM) < 1400

and The Notched Izod (NI) value at 73 degrees F in ft-lb/in needs to be in the range of 0.4 <= (NI) < 50

and The Elongation at Break (EB) value in percentage needs to be in the range of 3 <= (EB) < 400 or 5 <= (EB) < 400 or 7 <= (EB) < 400 or 10 <= (EB) < 400 or 25 <= (EB) < 400

and The Creep Modulus (CM) value in K p.s.i. needs to be in the range of 35 <= (CM) < 1400 or 5 <= (CM) < 1400 or 75 <= (CM) < 1400 or 150 <= (CM) < 1400 or 250 <= (CM) < 1400

and The Dielectric Strength (DS) value in V/mil needs to be in the range of 350 <= (DS) < 600 or 400 <= (DS) < 600 or 450 <= (DS) < 600 or 500 <= (DS) < 600

and The Arc Resistance (AR) value in sec. needs to be in the range of 5 <= (AR) < 250 or 50 <= (AR) < 250 or 75 <= (AR) < 250 or 100 <= (AR) < 250 or 125 <= (AR) < 250 or 150 <= (AR) < 250

THEN:

[ACETAL] IS GIVEN THE VALUE 2

NOTE:

This rule will evaluate whether ACETAL is an OKAY choice based on its price, maximum end-use temperature, anticipated environment and structural requirements in use.

REFERENCE:

Charles P. MacDermott and Aroon V. Shenoy, Selecting Thermoplastics for Engineering Applications, Marcel Dekker Inc., New York, 1997 (Tables in Chapter 11)

RULE NUMBER: 14

IF:

```
      [ACETAL] = 2
  and [P_ACETAL] < [ECP]
```

THEN:

```
      Acetal - Probability = 71/100
```

NOTE:

P_ACETAL = 6.3 is the Price of ACETAL in cents per cubic inch and ECP is the Expected Cost of Product in cents per cubic inch. It is ranked no. 5 for Price.

REFERENCE:

Charles P. MacDermott and Aroon V. Shenoy, Selecting Thermoplastics for Engineering Applications, Marcel Dekker Inc., New York, 1997 (Table 11.14)

RULE NUMBER: 15

IF:

```
      [ACETAL] = 2
  and [MAX_ET] < [HDT_ACETAL]
```

THEN:

```
      Acetal - Probability = 50/100
```

NOTE:

MAX_ET is the MAXimum End-use Temperature in degrees Fahrenheit and HDT_ACETAL = 277 is the Heat Deflection Temperature of ACETAL at 264 p.s.i. in degrees Fahrenheit. It is ranked no. 8 for HDT.

REFERENCE:

Charles P. MacDermott and Aroon V. Shenoy, Selecting Thermoplastics for Engineering Applications, Marcel Dekker Inc., New York, 1997 (Table 11.6)

o
o
o
o
o

RULE NUMBER: 50

o
o
o
o
o

RULE NUMBER: 75

o
o
o

o
o
RULE NUMBER: 100
o
o
o
o
o
RULE NUMBER: 125
o
o
o
o
o
RULE NUMBER: 150
o
o
o
o
o
RULE NUMBER: 169

IF:
[P_PET30G] < [ECP]
and [MAX_ET] < [HDT_PET30G]
and The Engineering Thermoplastic product during end-use application is likely to be exposed to the following chemicals/solvents: NOT Ketone or Aldehyde
and The Engineering Thermoplastic product during end-use application is held at all times at normal temperatures of about 73 degrees Fahrenheit or held frequently at elevated temperatures for extended period of time or held frequently at temperatures below freezing (less than 32 degrees Fahrenheit) for extended period of time
and The Engineering Thermoplastic product during end-use application is at all times NOT immersed in water whose temperature exceeds 125 degrees Fahrenheit
and The ambient relative humidity (where the product is being manufactured and/or being used) is greater than 50% and changes frequently with time or greater than 50% and does not change with time or less than 50% and changes frequently with time or less than 50% and does not change with time
and The proposed application involves the use of metals and the thermoplastic as part of a single item: Yes or No
and The Tensile Strength (TS) value in K p.s.i. needs to be in the range of 4 <= (TS) < 23 or 7 <= (TS) < 23 or 10 <= (TS) < 23 or 15 <= (TS) < 23 or 19 <= (TS) < 23 or 21 <= (TS) < 23
and The Flexural Modulus (FM) value in K p.s.i. needs to be in the range of 100 <= (FM) < 1400 or 300 <= (FM) < 1400 or

400 <= (FM) < 1400 or 600 <= (FM) < 1400 or 800 <= (FM) < 1400 or 1000 <= (FM) < 1400 or 1200 <= (FM) < 1400
and The Notched Izod (NI) value at 73 degrees F in ft-lb/in needs to be in the range of 0.4 <= (NI) < 50 or 1.5 <= (NI) < 50
and The Elongation at Break (EB) value in percentage needs to be in the range of 3 <= (EB) < 400
and The Creep Modulus (CM) value in K p.s.i. needs to be in the range of 35 <= (CM) < 1400 or 50 <= (CM) < 1400 or 75 <= (CM) < 1400 or 150 <= (CM) < 1400 or 250 <= (CM) < 1400 or 350 <= (CM) < 1400 or 450 <= (CM) < 1400 or 550 <= (CM) < 1400 or 650 <= (CM) < 1400 or 750 <= (CM) < 1400 or 1000 <= (CM) < 1400
and The Dielectric Strength (DS) value in V/mil needs to be in the range of 350 <= (DS) < 600 or 400 <= (DS) < 600 or 450 <= (DS) < 600 or 500 <= (DS) < 600 or 550 <= (DS) < 600
and The Arc Resistance (AR) value in sec. needs to be in the range of 5 <= (AR) < 250 or 50 <= (AR) < 250 or 75 <= (AR) < 250 or 100 <= (AR) < 250 or 125 <= (AR) < 250
and Good product appearance and gloss is not of critical importance

THEN:

[PET30G] IS GIVEN THE VALUE 15

NOTE:
This rule will evaluate whether Polyethylene terephthalate (30% glass) is an OKAY choice based on its price, maximum end-use temperature, anticipated environment and structural requirements in use and product aesthetic requirements.

REFERENCE:
Charles P. MacDermott and Aroon V. Shenoy, Selecting Thermoplastics for Engineering Applications, Marcel Dekker Inc., New York, 1997 (Tables in Chapter 11)

RULE NUMBER: 170

IF:

[PET30G] = 15
and [P_PET30G] < [ECP]

THEN:

Polyethylene terephthalate (30% glass) - Probability = 43/100

NOTE:
P_PET30G = 7.4 is the Price of Polyethylene terephthalate (30% glass) in cents per cubic inch and ECP is the Expected Cost of Product in cents per cubic inch. It is ranked no. 9 for Price.

REFERENCE:
Charles P. MacDermott and Aroon V. Shenoy, Selecting Thermoplastics for Engineering Applications, Marcel Dekker Inc., New York, 1997 (Table 11.14)

RULE NUMBER: 171

IF:
[PET30G] = 15
and [MAX_ET] < [HDT_PET30G]

THEN:
Polyethylene terephthalate (30% glass) - Probability = 85/100

NOTE:
MAX_ET is the MAXimum End-use Temperature in degrees Fahrenheit and HDT_PET30G = 435 is the Heat Deflection Temperature of Polyethylene terephthalate (30% glass) at 264 p.s.i. in degrees Fahrenheit. It is ranked no. 3 for HDT.

REFERENCE:
Charles P. MacDermott and Aroon V. Shenoy, Selecting Thermoplastics for Engineering Applications, Marcel Dekker Inc., New York, 1997 (Table 11.6)

RULE NUMBER: 172

IF:
[PET30G] = 15
and The Engineering Thermoplastic product during end-use application is likely to be exposed to the following chemicals/solvents: Strong acid or Gas or Alcohol or None of the above or Unknown

THEN:
Polyethylene terephthalate (30% glass) - Probability = 100/100

NOTE:
Polyethylene terephthalate (30% glass) is not affected by Strong acid or Gas or Alcohol at room temperature.

REFERENCE:
Charles P. MacDermott and Aroon V. Shenoy, Selecting Thermoplastics for Engineering Applications, Marcel Dekker Inc., New York, 1997 (Table 11.2)

RULE NUMBER: 173

IF:

[PET30G] = 15

and The Engineering Thermoplastic product during end-use application is likely to be exposed to the following chemicals/solvents: Strong base or Salt solution or Aromatics or None of the above or Unknown

THEN:

Polyethylene terephthalate (30% glass) - Probability = 50/100

NOTE:

Polyethylene terephthalate (30% glass) is slightly affected by Strong base or Salt solution or Aromatics at room temperature.

REFERENCE:

Charles P. MacDermott and Aroon V. Shenoy, Selecting Thermoplastics for Engineering Applications, Marcel Dekker Inc., New York, 1997 (Table 11.2)

RULE NUMBER: 174

IF:

[PET30G] = 15

and The Tensile Strength (TS) value in K p.s.i. needs to be in the range of 4 <= (TS) < 23 or 7 <= (TS) < 23 or 10 <= (TS) < 23 or 15 <= (TS) < 23 or 19 <= (TS) < 23 or 21 <= (TS) < 23

THEN:

Polyethylene terephthalate (30% glass) - Probability = 99/100

NOTE:

Tensile Strength is to be measured under ASTM D-638. (TS) value of less than 5 K p.s.i. is not normally considered for most engineering thermoplastic applications. (TS) value for Polyethylene terephthalate (30% glass) = 22.3 K p.s.i. It is ranked no. 1 for this property.

REFERENCE:

Charles P. MacDermott and Aroon V. Shenoy, Selecting Thermoplastics for Engineering Applications, Marcel Dekker Inc., New York, 1997 (Table 11.3)

RULE NUMBER: 175

IF:

[PET30G] = 15

and The Flexural Modulus (FM) value in K p.s.i. needs to be in the range of 100 <= (FM) < 1400 or 300 <= (FM) < 1400 or

400 <= (FM) < 1400 or 600 <= (FM) < 1400 or 800 <= (FM) < 1400 or 1000 <= (FM) < 1400 or 1200 <= (FM) < 1400

THEN:

Polyethylene terephthalate (30% glass) - Probability = 99/100

NOTE:
Flexural Modulus is to be measured under ASTM D-790. (FM) value for Polyethylene terephthalate (30% glass) = 1300 K p.s.i. It is ranked no. 1 for this property.

REFERENCE:
Charles P. MacDermott and Aroon V. Shenoy, Selecting Thermoplastics for Engineering Applications, Marcel Dekker Inc., New York, 1997 (Table 11.4)

RULE NUMBER: 176

IF:

[PET30G] = 15

and The Notched Izod (NI) value at 73 degrees F in ft-lb/in needs to be in the range of 0.4 <= (NI) < 50 or 1.5 <= (NI) < 50

THEN:

Polyethylene terephthalate (30% glass) - Probability = 36/100

NOTE:
Notched Izod at 73 degrees F is to be measured under ASTM D 256. An Izod over 1.5 represents a resin that is reasonably tough. (NI) value for Polyethylene terephthalate (30% glass) = 1.9 ft-lb/in. It is ranked no. 10 for this property.

REFERENCE:
Charles P. MacDermott and Aroon V. Shenoy, Selecting Thermoplastics for Engineering Applications, Marcel Dekker Inc., New York, 1997 (Table 11.7)

RULE NUMBER: 177

IF:

[PET30G] = 15

and The Elongation at Break (EB) value in percentage needs to be in the range of 3 <= (EB) < 400

THEN:

Polyethylene terephthalate (30% glass) - Probability = 8/100

NOTE:
Percentage Elongation at Break is to be measured under ASTM D-638. (EB) value for Polyethylene terephthalate (30% glass) = 3%. It is ranked no. 14 for this property.

REFERENCE:
Charles P. MacDermott and Aroon V. Shenoy, Selecting Thermoplastics for Engineering Applications, Marcel Dekker Inc., New York, 1997 (Table 11.8)

RULE NUMBER: 178

IF:

[PET30G] = 15

and The Creep Modulus (CM) value in K p.s.i. needs to be in the range of 35 <= (CM) < 1400 or 50 <= (CM) < 1400 or 75 <= (CM) < 1400 or 150 <= (CM) < 1400 or 250 <= (CM) < 1400 or 350 <= (CM) < 1400 or 450 <= (CM) < 1400 or 550 <= (CM) < 1400 or 650 <= (CM) < 1400 or 750 <= (CM) < 1400 or 1000 <= (CM) < 1400

THEN:

Polyethylene terephthalate (30% glass) - Probability = 92/100

NOTE:
(CM) value for Polyethylene terephthalate (30% glass) = 1000 K p.s.i. It is ranked no. 2 for this property.

REFERENCE:
Charles P. MacDermott and Aroon V. Shenoy, Selecting Thermoplastics for Engineering Applications, Marcel Dekker Inc., New York, 1997 (Table 11.9)

RULE NUMBER: 179

IF:

[PET30G] = 15

and The Dielectric Strength (DS) value in V/mil needs to be in the range of 350 <= (DS) < 600 or 400 <= (DS) < 600 or 450 <= (DS) < 600 or 500 <= (DS) < 600 or 550 <= (DS) < 600

THEN:

Polyethylene terephthalate (30% glass) - Probability = 99/100

NOTE:
Dielectric Strength is to be measured under ASTM D-149. (DS) value for Polyethylene terephthalate (30% glass) = 550 V/mil. It is ranked no. 1 for this property.

REFERENCE:
Charles P. MacDermott and Aroon V. Shenoy, Selecting Thermoplastics for Engineering Applications, Marcel Dekker Inc., New York, 1997 (Table 11.10)

RULE NUMBER: 180

IF:

[PET30G] = 15

and The Arc Resistance (AR) value in sec. needs to be in the range of 5 <= (AR) < 250 or 50 <= (AR) < 250 or 75 <= (AR) < 250 or 100 <= (AR) < 250 or 125 <= (AR) < 250

THEN:

Polyethylene terephthalate (30% glass) - Probability = 71/100

NOTE:
Arc Resistance is to be measured under ASTM D-495. (AR) value for Polyethylene terephthalate (30% glass) = 135 sec. It is ranked no. 5 for this property.

REFERENCE:
Charles P. MacDermott and Aroon V. Shenoy, Selecting Thermoplastics for Engineering Applications, Marcel Dekker Inc., New York, 1997 (Table 11.11)

RULE NUMBER: 181

IF:

The Engineering Thermoplastic product during end-use application is held frequently at elevated temperatures for extended period of time

THEN:

If the product is held frequently at elevated temperatures for extended period of time, then please note that heat ageing can cause eventual deterioration of polymer properties or even complete failure of its structure. It is important to bear this point in mind during resin selection.

NOTE:
WARNING MESSAGE

REFERENCE:
Charles P. MacDermott and Aroon V. Shenoy, Selecting Thermoplastics for Engineering Applications, Marcel Dekker Inc., New York, 1997 (Chapter 2)

RULE NUMBER: 182

IF:

The Engineering Thermoplastic product during end-use application is held frequently at temperatures below freezing (less than 32 degrees Fahrenheit) for extended period of time

THEN:

If the product is held frequently below freezing (less than 32 degrees Fahrenheit) for extended period of time, then please note that impact resistance and toughness are reduced drastically even by a drop of few degrees in temperature in this range. It is important to bear this point in mind during resin selection.

NOTE:
WARNING MESSAGE

REFERENCE:
Charles P. MacDermott and Aroon V. Shenoy, Selecting Thermoplastics for Engineering Applications, Marcel Dekker Inc., New York, 1997 (Chapter 2)

RULE NUMBER: 183

IF:

The proposed application involves the use of metals and the thermoplastic as part of a single item: Yes

THEN:

If the proposed application involves the use of metals and the thermoplastic as part of a single item, then please note that care has to be taken to allow for differences in the coefficient of thermal expansion values between the two, when there is likelihood of temperature changes. It is important to bear this point in mind during resin selection and keep away from unreinforced resins if possible.

NOTE:
WARNING MESSAGE

REFERENCE:
Charles P. MacDermott and Aroon V. Shenoy, Selecting Thermoplastics for Engineering Applications, Marcel Dekker Inc., New York, 1997 (Chapter 2)

RULE NUMBER: 184

IF:

```
      [ABS] = 0
  and [ACETAL] = 0
  and [6NYLON] = 0
  and [66NYLON] = 0
  and [66NYLON30G] = 0
  and [66NYLON_MR] = 0
  and [66NYLON_ST] = 0
  and [PC] = 0
  and [PC30G] = 0
```

```
  and  [PPE] = 0
  and  [PPE30G] = 0
  and  [PP40T] = 0
  and  [PS] = 0
  and  [PBT30G] = 0
  and  [PET30G] = 0
THEN:
       The resin selection process has failed to give you a choice.
       probably the anticipated environmental and structural
       requirements are too demanding and restrictive. Kindly relax
       one or more of the requirements and rerun the program to see
       if it does give a possible list of choices for resin
       selection.
NOTE:
ADVISORY MESSAGE
REFERENCE:
Charles P. MacDermott and Aroon V. Shenoy, Selecting Thermoplastics
for Engineering Applications, Marcel Dekker Inc., New York, 1997
```

EXAMPLE SESSION WITH SELECTHER

The concepts discussed in the preceding sections can be best illustrated by an actual sample run of SELECTHER through a screen-by-screen demonstration. A WORKING program of SELECTHER has to be made available first. This requires 13 files as listed in Appendix B along with the COMMAND.COM file to be present on a hard disk, 3.5-in. diskette, or 5.25 in. diskette.

Once a WORKING program is available on a diskette or hard disk, a sample run can be tried as follows. The transcript essentially shows the interactions with the user in terms of questions, recommendations, and explanations.

The actual user input is given in bold. {added comments and annotations are within curly parentheses}, while the text output from SELECTHER is the remainder.

The program begins when the user types the word SELECTA at the DOS prompt on an IBM PC or compatible computer.

A>**SELECTA**

SCREEN 1 {The processing of Reading Rules takes different lengths of time depending on the central processing unit speed.}

```
                    E X S Y S

          Expert System Runtime Program

     (c) copyright 1983,84,85 EXSYS, INC.

                  Ver.3.2.12


                 READING RULES
```

Screen 2 {Upon pressing any key, the program asks the following.}

Recover previously saved position Y/N (Default = N):

{Answer Y only to retrieve data that has been saved during an earlier consultation using the Quit command; or else, press ENTER to choose the N default action.}

S E L E C T H E R

Expert System Demonstration Program for Selecting Thermoplastics for Engineering Applications

by: SHENOY

Press any key to start:

Screen 3 {Inputting **25** and pressing <ENTER> brings up SCREEN 4.}

Please input Estimated Cost of Product in cents per cubic inch
: 25

Input a value for the variable between 1.000000 and 250.000000
'WHY' for information, <?> for more details or 'QUIT' to save data:

SCREEN 4 {Inputting **100** and pressing <ENTER> brings up SCREEN 5.}

```
Please input Maximum End-use Temperature in degrees Fahrenheit
: 100

Input a value for the variable between 70.000000 and 600.000000
'WHY' for information, <?> for more details or 'QUIT' to save
data:
```

Screen 5 {Pressing <H> will project SCREEN 6 and provide some explanations.}

```
The Engineering Thermoplastic product during end-use application
is likely to be exposed to the following chemicals/solvents:
     1    Strong acid
     2    Strong base
     3    Salt solution
     4    Aromatics
     5    Gas
     6    Alcohol
     7    Ketone
     8    Aldehyde
     9    None of the above
     10   Unknown
 H

Enter number(s) of value(s), WHY for information on the rule,
<?> for more details, QUIT to save data entered or <H> for help
```

Screen 6 {Pressing <SPACE> will return SCREEN 5, redisplayed as SCREEN 7.}

A V S **SELECTHER** Help

SELECTHER is basically asking you for a numerical input in order to proceed with the consultation. When the numbered list has more than one item, input the relevant number(s) separated by a comma or space and then press the <ENTER> key to make the selection. If there is just a single item, then input the number 1 and press the <ENTER> key to continue.

If you do not understand why **SELECTHER** is asking you this question, you can ask it what rule it is trying to apply by typing WHY and then pressing the <ENTER> key. **SELECTHER** will respond by displaying the rule it is trying to validate. Pressing the <ENTER> key will re-display the original screen.

Typing <?> and pressing the <ENTER> key gets more details such as definitions of certain terms. However, this facility is not available for every screen.

If you want to store the data that you have input so far and exit **SELECTHER** enter QUIT in response to the request for information. When asked to name the file in which the data is to be stored, DO NOT USE the name **SELECTA** as the data will get stored in existing file and previous information will be erased.

TO RETURN TO PROGRAM PRESS <SPACE>

SCREEN 7 {Choosing WHY and pressing <ENTER> brings up SCREEN 8, which displays the Rule that SELECTHER is trying to execute.}

```
The Engineering Thermoplastic product during end-use application
is likely to be exposed to the following chemicals/solvents:
     1    Strong acid
     2    Strong base
     3    Salt solution
     4    Aromatics
     5    Gas
     6    Alcohol
     7    Ketone
     8    Aldehyde
     9    None of the above
     10   Unknown
 WHY

Enter number(s) of value(s), WHY for information on the rule,
<?> for more details, QUIT to save data entered or <H> for help
```

Screen 8 {Pressing <H> brings up SCREEN 9.}

{RULE NUMBER: 1 has been presented in detail in an earlier section and hence is not repeated here.}

```
RULE  NUMBER :  1

IF :

        (1)  [P_ABS] < [BCP]
and     (2)  [MAX_ET] < [HDT_ABS]
and     (3)  The Engineering Thermoplastic product during end-use
             application is likely to be exposed to the following
             chemicals/solvents:  NOT Aromatics
and     (4)
         .
         .
and     (14)
THEN :
             ABS  IS GIVEN THE VALUE 1
NOTE :  This rule will evaluate whether ABS is an OKAY choice
        based on its price, maximum end-use temperature,
        anticipated environment and structural requirements in
        use.

IF line # for derivation, <K> - known data, <C> - choices, <R> - references
↑ or ↓ - prev. or next rule, <J> - jump, <H> for help or <ENTER> to continue:
```

Screen 9 {Pressing <SPACE> will return Screen 8, and then pressing <ENTER> will show RULE NUMBER: 2 and then return SCREEN 5 or 7 re-displayed as SCREEN 10.}

A V S **SELECTHER** Help

If you wish to know why **SELECTHER** considers one of the IF conditions to be true, enter the line number of the IF condition. **SELECTHER** will respond by presenting the origin of the information:

1. If the information came from asking you for input, **SELECTHER** will let you know that "you told it" so.

2. If the condition that you are asking about has not been checked yet, **SELECTHER** will tell you that "it does not yet know if it is true".

3. If the statement is a mathematical expression, then **SELECTHER** will display the value of each variable in the expression. If you want to know how the value for a specific variable was determined, enter the variable line number. **SELECTHER** will explain how that value was got.

Note that **SELECTHER** will re-display the rule after each question.

Press the <ENTER> key when you have finished asking about the rule.

TO RETURN TO PROGRAM PRESS <SPACE>

SCREEN 10 {Choosing ? and pressing <ENTER> brings up SCREEN 11.}

The Engineering Thermoplastic product during end-use application
is likely to be exposed to the following chemicals Solvents:

1 Strong acid
2 Strong base
3 Salt solution
4 Aromatics
5 Gas
6 Alcohol
7 Ketone
8 Aldehyde
9 None of the above
10 Unknown

?

Enter number(s) of value(s), WHY for information on the rule,
<?> for more details, QUIT to save data entered or <H> for help

Screen 11 {Pressing <SPACE> will return SCREEN 10, redisplayed as SCREEN 12.}

```
In case you would like to guess the possible chemical environment
that the thermoplastic would be exposed to, then you may choose
any one or more of the proposed eight chemicals/solvents.Choosing
none of the above or unknown will assume that the thermoplastic
is not affected by any of the chemicals/solvents listed.

TO RETURN TO PROGRAM  PRESS <SPACE>
```

Screen 12 {Choosing 9 and pressing <ENTER> brings up SCREEN 13.}

```
The Engineering Thermoplastic product during end-use application
is likely to be exposed to the following chemicals/solvents:
     1    Strong acid
     2    Strong base
     3    Salt solution
     4    Aromatics
     5    Gas
     6    Alcohol
     7    Ketone
     8    Aldehyde
     9    None of the above
     10   Unknown
 9

Enter number(s) of value(s), WHY for information on the rule,
<?> for more details, QUIT to save data entered or <H> for help
```

Screen 13 {Choosing 1 and pressing <ENTER> brings up SCREEN 14.}

```
The Engineering Thermoplastic product during end-use application
is
     1    held at all times at normal temperatures of about 73
          degrees Fahrenheit
     2    held frequently at elevated temperatures for extended
          period of time
     3    held frequently at temperatures below freezing (less
          than 32 degrees Fahrenheit) for extended period of time
1

______________________________________________________________
Enter number(s) of value(s), WHY for information on the rule
<?> for more details, QUIT to save data entered or <H> for help
```

Screen 14 {Choosing 3 and pressing <ENTER> brings up SCREEN 15.}

```
The Engineering Thermoplastic product during end-use application
is at all times
     1   immersed in water whose temperature exceeds 125 degrees
         Fahrenheit
     2   immersed in water whose temperature is normally at
         around 73 degrees Fahrenheit
     3   not immersed in water
3

Enter number(s) of value(s), WHY for information on the rule
<?> for more details, QUIT to save data entered or <H> for help
```

SCREEN 15 {Choosing 4 and pressing <ENTER> brings up SCREEN 16.}

```
The ambient relative humidity (where the product is being
manufactured and/or being used is
     1   greater than 50% and changes frequently with time
     2   greater than 50% and does not change with time
     3   less than 50% and changes frequently with time
     4   less than 50% and does not change with time
4

Enter number(s) of value(s), WHY for information on the rule
<?> for more details, QUIT to save data entered or <H> for help
```

SCREEN 16 {Choosing 2 and pressing <ENTER> brings up SCREEN 17.}

```
The proposed application involves the use of metals and the
thermoplastic as part of a single item:
     1   Yes
     2   No
2

Enter number(s) of value(s), WHY for information on the rule
<?> for more details, QUIT to save data entered or <H> for help
```

SCREEN 17 {Choosing 1 and pressing <ENTER> brings up SCREEN 18.}

```
The Tensile Strength (TS) value in K p.s.i. needs to be in the
range of
      1    4 <= (TS) < 23
      2    7 <= (TS) < 23
      3   10 <= (TS) < 23
      4   15 <= (TS) < 23
      5   19 <= (TS) < 23
      6   21 <= (TS) < 23
1
```

```
Enter number(s) of value(s), WHY for information on the rule
<?> for more details, QUIT to save data entered or <H> for help
```

Screen 18 {Choosing 1 and pressing <ENTER> brings up SCREEN 19.}

```
The Flexural Modulus (FM) value in K p.s.i. needs to be in the
range of
        1     100 <= (FM) < 1400
        2     300 <= (FM) < 1400
        3     400 <= (FM) < 1400
        4     600 <= (FM) < 1400
        5     800 <= (FM) < 1400
        6    1000 <= (FM) < 1400
        7    1200 <= (FM) < 1400
1

------------------------------------------------------------------

Enter number(s) of value(s), WHY for information on the rule
<?> for more details, QUIT to save data entered or <H> for help
```

Screen 19 {Choosing 1 and pressing <ENTER> brings up SCREEN 20.}

```
The Notched Izod (NI) value at 73 degrees F in ft-lb/in needs to
be in the range of
     1   0.4 <= (NI) < 50
     2   1.5 <= (NI) < 50
     3     3 <= (NI) < 50
     4    10 <= (NI) < 50
     5    25 <= (NI) < 50
1

Enter number(s) of value(s), WHY for information on the rule
<?> for more details, QUIT to save data entered or <H> for help
```

SCREEN 20 {Choosing 1 and pressing <ENTER> brings up SCREEN 21.}

```
The Elongation at Break (EB) value in percentage needs to be in
the range of
     1      3 <= (EB) < 400
     2      5 <= (EB) < 400
     3      7 <= (EB) < 400
     4     10 <= (EB) < 400
     5     25 <= (EB) < 400
     6     75 <= (EB) < 400
     7    100 <= (EB) < 400
     8    200 <= (EB) < 400
     9    300 <= (EB) < 400
1

Enter number(s) of value(s), WHY for information on the rule
<?> for more details, QUIT to save data entered or <H> for help
```

SCREEN 21 {Choosing 1 and pressing <ENTER> brings up SCREEN 22.}

```
The Creep Modulus (CM) value in K p.s.i. needs to be in the
range of
     1      35<= (CM)< 1400
     2      50<= (CM)< 1400
     3      75<= (CM)< 1400
     4     150<= (CM)< 1400
     5     250<= (CM)< 1400
     6     350<= (CM)< 1400
     7     450<= (CM)< 1400
     8     550<= (CM)< 1400
     9     650<= (CM)< 1400
     10    750<= (CM)< 1400
     11   1000<= (CM)< 1400
     12   1200<= (CM)< 1400
1

Enter number(s) of value(s), WHY for information on the rule
<?> for more details, QUIT to save data entered or <H> for help
```

SCREEN 22 {Choosing 1 and pressing <ENTER> brings up SCREEN 23.}

```
The Dielectric Strength (DS) value in V/mil needs to be in the
range of
      1   350 <= (DS) < 600
      2   400 <= (DS) < 600
      3   450 <= (DS) < 600
      4   500 <= (DS) < 600
      5   550 <= (DS) < 600
1

Enter number(s) of value(s), WHY for information on the rule
<?> for more details, QUIT to save data entered or <H> for help
```

Screen 23 {Choosing 1 and pressing <ENTER> brings up SCREEN 24.}

```
The Arc Resistance (AR) value in sec. needs to be in the range of
      1      5 <= (AR) < 250
      2     50 <= (AR) < 250
      3     75 <= (AR) < 250
      4    100 <= (AR) < 250
      5    125 <= (AR) < 250
      6    150 <= (AR) < 250
1

Enter number(s) of value(s), WHY for information on the rule
<?> for more details, QUIT to save data entered or <H> for help
```

Screen 24 {Choosing 2 and pressing <ENTER> prompts SELECTHER to work through the rules, sort and print results, and display SCREEN 25.}

```
Good product appearance and gloss is
     1   of critical importance
     2   not of critical importance
2

Enter number(s) of value(s), WHY for information on the rule
<?> for more details, QUIT to save data entered or <H> for help
```

SCREEN 25 {Choosing <H> and pressing <ENTER> brings up SCREEN 26.}

Values based on - 100 to + 100 system

		VALUE
1	Polyethylene terephthalate (30% glass)	71
2	Polybutylene terephthalate (30% glass)	62
3	6/6 nylon (30% glass)	60
4	Acetal	59
5	PPE-based resin	59
6	6/6 nylon (mineral reinforced)	58
7	PPE-based resin (30% glass	58
8	Polycarbonate (30% glass)	55
9	6/6 nylon	50
10	Polypropylene (40% talc)	50
11	Polysulfone	50
12	ABS	48
13	Polycarbonate	47
14	6 nylon	44
15	6/6 nylon (Super Tough)	41

All choices <A>, only if value > 1 <G>, Print <P>, Change and rerun <C>,
New sort type <S>, rules used <line #>, Quit/save <Q>, Help <H>, Done <D>:H

SCREEN 26 {Pressing <SPACE> returns SCREEN 25, redisplayed as SCREEN 27.}

```
A V S                          SELECTHER                              Help
--------------------------------------------------------------------------
SELECTHER has displayed the  RESINS  along with their certainty factors based
on requirements that were input by you.  You may now ask SELECTHER about the
rule(s) it fired to arrive at a particular answer by entering the line number.

Pressing <C> will display all inputs that SELECTHER acquired.   By entering a
line number, you can alter the data.    When the data is in the form you wish,
you may press <R> to rerun.

Pressing <P> will allow you to print the output screen in  standard format  or
in the report generator form.   In the standard format, you have the option of
also printing all the input that was given to SELECTHER.

Pressing <Q> will allow you to store data and exit SELECTHER.  When asked for
the filename to store the data,  DO NOT use the name  SELECTA as the original
information in the existing file will be erased during data storage.

Pressing <A> or <S> will re-display the list; pressing <D> ends consultation.

--------------------------------------------------------------------------
TO RETURN TO PROGRAM PRESS <SPACE>
```

Screen 27 {Note that the resins are listed in prioritized order based on the general usefulness ranking as given in Table 11.17, since the answers to the questions were not restrictive in any manner.}
{Choosing P and pressing <ENTER> gives an immediate printout of the final screen in the standard format. The Report of the consultation is also saved in a file named REPORT, which can be printed out at a later time. However, it is to be noted that the old file REPORT is deleted before every fresh consultation and a new file REPORT is created. Choosing D and pressing <ENTER> ends the consultation but gives the following option.}

Run again (Y/N)(Default = N):

{"Run again" option is useful in carrying out successive consultations without program reloading. For instance, you may rerun the program for the two examples given in Chapter 11.}
{Choosing the N default option will lead to the DOS prompt. Upon using the command TYPE REPORT, the Report given in Appendix B is displayed on the screen.}

	Values based on - 100 to + 100 system	VALUE
1	**Polyethylene terephthalate (30% glass)**	**71**
2	**Polybutylene terephthalate (30% glass)**	**62**
3	**6/6 nylon (30% glass)**	**60**
4	**Acetal**	**59**
5	**PPE-based resin**	**59**
6	**6/6 nylon (mineral reinforced)**	**58**
7	**PPE-based resin (30% glass)**	**58**
8	**Polycarbonate (30% glass)**	**55**
9	**6/6 nylon**	**50**
10	**Polypropylene (40% talc)**	**50**
11	**Polysulfone**	**50**
12	**ABS**	**48**
13	**Polycarbonate**	**47**
14	**6 nylon**	**44**
15	**6/6 nylon (Super Tough)**	**41**

All choices <A>, only if value > 1 <G>, Print <P>, Change and rerun <C>,
New sort type <S>, rules used <line #>, Quit/save <Q>, Help <H>, Done <D>:D

References

1. Shenoy, A. V., Saini, D. R., and Nadkarni, V. M., Rheograms for engineering thermoplastics from melt flow index, *Rheol. Acta*, 22: 209–222 (1983).
2. Bonfield, W., Grynpas, M. D., Tully, A. E., Bowman, J., and Abram, J., Hydroxyapatite reinforced polyethylene—A mechanically compatible implant material for bone replacement, *Biomaterials*, 2: 185–186 (1981).
3. Copeland, J. R., and Rush, O. W., Reinforcing with wollastonite filler makes for a tougher polypropylene, *Mod. Plastics*, 56: 68–69 (1979).
4. Xanthos, M., MICA: Filler/reinforcement in flake form, *Plastics Compounding* (July/Aug. 1979).
5. Balow, M. J., and Fuccella, D. C., Hybridization of reinforcement to optimize part performance and molding in reinforced thermoplastics, *Polymer-Plastic Technol. Eng.*, 20; 23–33 (1983).
6. Copeland, J. R., and Rush, O. W., Wollastonite: Short-fiber filler/reinforcement, *Plastics Compounding*, 1: 26–35 (1978).

7. Griffin, G. J. L., Orientation effects in filled plastics bearing materials, *ASLE Trans.*, 15: 171–176 (1972).
8. Runt, J., and Galgoci, E. C., Polymer/piezoelectric ceramic composites: Polystyrene and poly(methyl methacrylate) with PZT, *J. Appl. Polymer Sci.*, 29; 611–617 (1984).
9. Runt, J., and Galgoci, E. C., Piezoelectric composites of PZT and some semicrystalline polymers, *Mater. Res. Bull.*, 19: 253–260 (1984).
10. Norman, J. C., Kevlar aramid fiber developments in reinforced plastics and friction materials, Paper presented at Society of Advancement of Material and Process Engineering Seminar, Southfield, MI, Nov. 15, 16, 1978.
11. Hancox, N. L., and Minty, D. C. C., Materials qualification and property measurements of carbon fibre reinforced composites for space use, *J. Br. Interplanet. Soc.*, 30: 391–399 (1977).
12. Davenport, D. E., Metalloplastics: An answer to electromagnetic pollution, Paper presented at Conference of the Society of Plastics Engineers, Cleveland, OH Nov. 18, 1980.
13. Warfel, R. H., Metallized glass fiber: The long and short conductive plastics, Paper presented at 36th Annual Conference, Reinforced Plastics/Composites Institute, Society of the Plastics Industry, Feb. 16–20, 1981.
14. Saini, D. R., Shenoy, A. V., and Nadkarni, V. M., Melt rheology of highly loaded ferrite-filled polymer composites, *Polymer Compos.*, 7: 193–200 (1986).
15. Saini, D. R., and Shenoy, A. V., Viscoelastic properties of highly loaded ferrite-filled polymer systems, *Polymer Eng. Sci.*, 26: 441–445 (1986).
16. Katz, H. S., and Milewski, J. V., *Handbook of Fillers and Reinforcements for Plastics*, Van Nostrand Reinhold, New York (1978).
17. Mascia, L., *The Role of Additives in Plastics*, Edward Arnold, London (1974).
18. Deanin, R. D., and Schott, N. R., *Fillers and Reinforcements for Plastics*, Adv. Chem. Ser. 134, American Chemical Society, Washington, DC (1974).
19. Wake, W. C., *Fillers for Plastics*, Illiffe Books, London (1971).
20. Richardson, M. O. W., *Polymer Engineering Composites*, Applied Science Publishers, London (1977).
21. Mallick, P. K., *Fiber-Reinforced Composites: Materials, Manufacturing, and Design*, Marcel Dekker, New York (1993).
22. Shenoy, A. V., Rheology of highly filled polymer melt systems, in *Encyclopedia of Fluid Mechanics* (N. P. Cheremisinoff, ed.), Vol. 7, pp. 667–701, Gulf Publishing, Houston, TX (1988).
23. Smeykal, J. P., *Modern Plastics Encyclopedia*, Vol. 126, McGraw-Hill, New York (1979–80).
24. Fried, J. R., Polymer technology—Part 7: Engineering thermoplastics and specialty plastics, *Plastics Eng.*, pp. 35–44 (May 1983).
25. Saini, D. R., and Shenoy, A. V., Melt rheology of some specialty polymers, *J. Elastomers Plastics*, 17: 189–217 (1985).
26. Mandelkern, L., *Crystallization of Polymers*, McGraw-Hill, New York (1964).
27. Meares, P., *Polymers: Structure and Bulk Properties*, Van Nostrand Reinhold, New York (1965).
28. Tager, A., *Physical Chemistry of Polymers*, Mir, Moscow (1978).

29. Bergenn, W. R., and Rigby, R. B., PES and PEEK: Tough engineering thermoplastics, *Chem. Eng. Prog.*, 36–38 (Jan. 1985).
30. *Plastics Design Forum*, p. 70 (May–June 1979).
31. Sneller, J. A., and Smoluk, G. R., Special report on plastics in the medical market: Where engineering resins fit and what they do, *Mod. Plastics Int.*, pp. 36–45 (May 1984).
32. Churma, J. L., and Chapman, R. D., Nylon properties and applications, *Chem. Eng. Prog.*, pp. 49–54 (Jan. 1985).
33. Dix, J. S., PPS: The versatile engineering plastic, *Chem. Eng. Prog.*, pp. 42–44 (Jan 1985).
34. Wendle, B. C., *What Every Engineer Should Know About Developing Plastics Products*, Marcel Dekker, New York (1991).
35. *Modern Plastcs Encyclopedia*, Vol. 56, No. 10a, pp. 489–494, McGraw-Hill, New York (1979–80).
36. Kambour, R. P., Environmental stress cracking of thermoplastics, *Corrosion Fatigue*, Nace 2, pp. 681–684 (1972).
37. Deanin, R. D., and Hauser, D. I., Recent developments in environmental stress-crack resistance of plastics, *Polymer-Plastic Technol. Eng.*, 17: 123–137 (1981).
38. Tsuji, K., ESR study of photodegradation of polymers, *Polymer-Plastic Technol. Eng.*, 9: 1–86 (1977).
39. Mai, Y. W., Head, D. R., Cotterell, B., and Roberts, B. W., Mechanical properties of nylon 6 subjected to photodegradation, *J. Mater. Sci.*, 15: 3057–3065 (1980).
40. Wiles, D. M., and Carlsson, D. J., Stop photodegradation, *Chem. Tech.*, 11: 158–161 (1981).
41. Paolino, P. R., Antioxidants: Inhibiting polymer degradation, *Plastics Compounding* (Sept./Oct. 1980).
42. Gugumus, F., Developments in the U.V. stabilisation of polymers, in *Developments in Polymer Stabilisation 1* (G. Scott, ed.), pp. 261–308, Applied Science Publishers, Essex, England (1979).
43. Martin, J. R., and Gardner, R. J., Effect of long term humid aging in plastics, *Polymer Eng. Sci.*, 21: 557–565 (1981).
44. Heijboer, J., in *Static and Dynamic Properties of the Polymeric Solid State* (R. A. Pethrick and R. W. Richards, eds.), p. 197, (1982).
45. Murayama, T., *Dynamic Mechanical Analysis of Polymeric Material*, Elsevier, Amsterdam, Oxford, New York (1978).
46. Read, B. E., and Dean, G. D., *The Determination of Dynamic Properties of Polymers and Composites*, John Wiley, New York (1978).
47. Shenoy, A. V., and Saini, D. R., One-day test to predict long term mechanical behaviour of plastics for a year, *Polymer Testing*, 6: 37–45 (1986).
48. Saini, D. R., and Shenoy, A. V., Can the ASTM heat distortion test provide more than just a single-point measurement?, *Polymer Testing*, 7: 293–297 (1987).
49. O'Brien, K. T., Wear control in polymer processing, *Polymer-Plastic Technol. Eng.*, 18: 149–166 (1982).
50. Olmstead, B. A., How glass-fiber fillers affect injection machines, *SPE J.*, 26: 42–43 (1970).
51. Boothroyd, G., and Dewhurst, P., *Machine Design* (Jan. 26, 1984).

52. Shenoy, A., and Shenoy, U., Expert systems in plastics processing, *Materials Eng.*, 105: 33–36 (Nov. 1988).
53. Shenoy, A. V., and Shenoy, U. V., Extrudoc—A learning expert system to rectify extrusion defects, *Plastics Rubber Int.*, 15: 22–28 (Aug. 1990).
54. Shenoy, A. V., Practical applications of rheology to polymer processing, in Encyclopedia of Fluid Mechanics (N. P. Cheremisinoff, ed.), Vol. 7, pp. 961–989, Gulf Publishing, Houston, TX (1988).
55. American Society for Testing Materials, Philadelphia, 1979.
56. Beck, R. D., *Plastic Product Design*, Van Nostrand Reinhold, New York (1968).
57. Richardson, P. N., A guide to standard physical tests for plastics, Report from E. I. du Pont de Nemours & Co., Wilmington, DE (April 1962).
58. Ogorkiewicz, M., ed., *Thermoplastic Properties and Design*, pp. 38, 56, 234, John Wiley, New York (1983).
59. Powell, P. C., *Engineering with Polymers*, p. 77, Chapman & Hall, New York (1983).
60. Saini, D. R., and Shenoy, A. V., Deformation behavior of poly(vinylidene fluoride), *Ind. Eng. Chem. Prod. Res. Dev.*, 25: 277–282 (1986).
61. Shenoy, A. V., and Saini, D. R., *Thermoplastic Melt Rheology and Processing*, Marcel Dekker, New York (1996).
62. Harper, C. A., What you should know about plastics processing, *Chem. Eng.*, pp. 100–114 (May 10, 1976).
63. Flow Rates of Thermoplastics by Extrusion Plastimeter, ASTM D1238, American Society for Testing Materials, Philadelphia (1979).
64. Shida, M., Shroff, R. N., and Cancio, L. V., Correlation of low density polyethylene rheological measurements with optical and processing properties, *Polymer Eng. Sci.*, 17: 769–774 (1977).
65. Smith, D. J., The correlation of melt index and extrusion coating resin performance, *TAPPI*, 60: 131–133 (1977).
66. Shenoy, A. V., Chattopadhyay, S., and Nadkarni, V. M., From melt flow index to rheogram, *Rheol. Acta*, 22: 90–101 (1983).
67. Saini, D. R., and Shenoy, A. V., Viscoelastic properties of linear low density polyethylene, *Eur. Polymer J.*, 19: 811–816 (1983).
68. Shenoy, A. V., Saini, D. R., and Nadkarni, V. M., Rheograms of cellulosic polymers from melt flow index, *J. Appl. Polymer Sci.*, 27: 4399–4408 (1982).
69. Shenoy, A. V., Saini, D. R., and Nadkarni, V. M., Rheology of poly(vinyl chloride) formulations from melt flow index measurements, *J. Vinyl Tech.*, 5: 192–197 (1983).
70. Shenoy, A. V., and Saini, D. R., Copolymer melt rheograms from melt flow index, *Br. Polymer J.*, 17: 314–322 (1985).
71. Shenoy, A. V., and Saini, D. R., Melt rheology of liquid crystalline polymer, *Mol. Cryst. Liq. Cryst.*, 135: 343–354 (1986).
72. Shenoy, A. V., Saini, D. R., and Nadkarni, V. M., Melt rheology of polymer blends from melt flow index, *Int. J. Polymer, Mat.*, 10: 213–235 (1984).
73. Shenoy, A. V., Saini, D. R., and Nadkarni, V. M., Rheograms of filled polymers from melt flow index, *Polymer Composites*, 4: 53–63 (1983).

74. Shenoy, A. V., Saini, D. R., and Nadkarni, V. M., Rheology of nylon 6 containing metal halides, *J. Mater. Sci.*, 18: 2149–2155 (1983).
75. Saini, D. R., and Shenoy, A. V., A new method for the determination of flow activation energy of polymer melts, *J. Macromol. Sci.*, B22: 437–449 (1983).
76. Shenoy, A. V., and Saini, D. R., Rheological models for unified curves for simplified design calculations in polymer processing, *Rheol. Acta*, 23: 368–377 (1984).
77. Shenoy, U. V., Bamane, S., and Shenoy, A. V., A general rheological model for polymer melts, Paper presented at 40th Canadian Chemical Engineering Conference, Halifax, N.S., Canada (1990).
78. Saini, D. R., and Shenoy, A. V., Simplified calculations for mould filling during non-isothermal flow of polymer melts, *Plastic Rubber Proc. Appl.*, 3: 175–180 (1983).
79. Shenoy, A. V., and Saini, D. R., Compression moulding of ultra high molecular weight polyethylene, *Plastic Rubber Proc. Appl.*, 5: 313–317 (1985).
80. Ray, A., and Shenoy, A. V., PVC calendering: A simplified prediction technique, *J. Appl. Polymer Sci.*, 30: 1–18 (1985).
81. Shenoy, A. V., and Saini, D. R., Prediction of pressure losses through typical die shapes based on a simple novel approach, *Polymer-Plastic Technol. Eng.*, 23: 169–183 (1984).
82. Shenoy, A. V., and Saini, D. R., A simplistic route to viscous heat estimations in polymer processing, *Polymer-Plastic Technol. Eng.*, 23: 34–68 (1984).
83. Shenoy, A. V., Estimation of compounding conditions and grade selections in the preparation of thermoplastic melt blends, *Polymer-Plastic Technol. Eng.*, 24: 27–41 (1985).
84. Speed, C. S., Formulating blends of LLDPE and LDPE to design better film, *Plastics Eng.*, p. 39 (July 1982).
85. Shenoy, A. V., and Saini, D. R., An approach to the estimation of polymer melt elasticity, *Rheol. Acta*, 23: 608–616 (1984).
86. Shenoy, A. V., and Saini, D. R., A simplified approach to the prediction of primary normal stress differences in polymer melts, *Chem. Eng. Commun.*, 28: 1–27 (1984).
87. Shenoy, A. V., and Saini, D. R., Estimation of melt elasticity of degraded polymer from melt flow index, *Polymer Degrad. Stability*, 11: 297–307 (1985).
88. Shenoy, A. V., and Saini, D. R., Re-analysis of extensional flow data of polymer melts, *Angew. Makro. Chemie*, 137: 77–81 (1985).
89. Shenoy, A. V., and Saini, D. R., A new shift factor for coalescing dynamic viscoelastic data of polymer melts, *Acts Polymerica*, 37: 504–507 (1986).
90. Rubin, I. I., *Injection Molding: Theory and Practice*, John Wiley, New York (1973).
91. Bernhardt, E. C. (ed.), *Processing of Thermoplastic Materials*, Van Nostrand Reinhold, New York (1959).
92. McKelvey, J. M., *Polymer Processing*, John Wiley, New York (1962).
93. Holmes-Walker, W. A., *Polymer Conversion*, Halstead, London (1975).
94. Tadmor, Z., and Gogos, C. G., Principles of Polymer Processing, John Wiley, New York (1979).

95. Thorne, J. L., *Plastics Process Engineering*, Marcel Dekker, New York (1979).
96. Fenner, R. T., *Principles of Polymer Processing*, Chemical Publishing Co., New York (1980).
97. Isayev, A. I., *Injection and Compression Molding Fundamentals*, Marcel Dekker, New York (1987).
98. Dealy, J. M., Energy conservation in plastics processing: A review, *Polymer Eng. Sci.*, 22: 528–535 (1982).
99. Booy, M. L., The influence of non-Newtonian flow on effective viscosity and channel efficiency in screw pumps, *Polymer Eng. Sci.*, 21: 93–99 (1981).
100. Barrie, I. T., An application of rheology to the injection moulding of large-area articles, *Plastics and Polymers*, 38: 47–51 (1970).
101. Dubois, J. H., and Pribble, W. J., *Plastics Mold Engineering Handbook*, 3rd ed., Van Nostrand Rheinhold Co. (1978).
102. Wang, K. K., A system approach to injection molding process, *Polymer-Plastic Technol. Eng.*, 14: 75–93 (1980).
103. *Plastics Technology*, Industry News Section, Vol. 26, No. 11, pp. 147–155, Bill Communications, New York (Oct. 1980).
104. *Concepts in Engineering Plastics*, E. I. du Pont de Nemours & Co., Wilmington, DE (May 1978).
105. *Plastics Technology, Resin Pricing Update*, 61–62, Bill Communications Inc., New York, March 1996, pp. 61–62.
106. *Modern Plastics Encyclopedia*, Chemical Resistance, Vol. 55, No. 10a, pp. 499–560, McGraw-Hill, New York (1978–79).
107. *Celanese Nylon—Glass Reinforced Nylon Bulletin N1B*, Celanese Plastics and Specialty Co., Chatham, NJ (July 1979).
108. *Modern Plastics Encyclopedia*, Vol. 55, No. 10a, pp. 470, McGraw-Hill, New York (1978–79).
109. Cloud, P. J., and Wolverton, M. A., Product shrinkage and warpage of reinforced and filled thermoplastics, *Plastics Technology*, 24(12): 107–113 (Nov. 1978).
110. Leidner, J., Recovery of the value from postconsumer plastics waste, *Polymer-Plastic Technol. Eng.*, 10: 199–215 (1978).
111. Leidner, J., *Plastics Waste—Recovery of Economic Value*, Marcel Dekker, New York (1981).
112. Calendine, R. H., Palmer, M., and Bramer, P. V., Unsaturated polyester resins based on reclaimed polyethylene terephthalate (PET) beverage bottles, Paper presented at Thirty-Fifth Annual Conference of Reinforced Plastics/Composites Institute, SPI (1980).
113. Bremer, W. P., Photodegradable polyethylene, *Polymer-Plastic Technol. Eng.*, 18: 137–148 (1982).
114. Cossais, J.-C., Measures being taken in France to recycle plastics wastes, *German Plastics*, 68 (5): 2–4 (1978) [translated from *Kunststoffe*, 68 (5); 266–269] (1978).
115. Emminger, H., The recycling of plastics wastes in the Federal German Republic, *German Plastics*, 68 (5): 5–9 (1978) [translated from *Kunststoffe*, 68 (5): 270–277] (1978).

116. Ranby, B., Recycling of plastics—Experience in the Scandinavian countries, *German Plastics*, 68 (5): 10–11 (1978) [translated from *Kunststoffe*, 68 (5): 278–280] (1978).
117. Ferguson, W. C., Trends and developments in the recycling of plastics—Experiences in the United Kingdom, Japan and the USA, *German Plastics*, 68 (5): 25–28 (1978) [translated from *Kunststoffe*, 68 (5): 302–307 (1978)].
118. Roff, W. J., and Scott, J. R. (eds.), *Handbook of Common Polymers*, CRC Press, Boca Raton, FL, Butterworth, London (1971).
119. Frados, J. (ed.), *Plastics Engineering Handbook*, SPE 4th ed., Van Nostrand Reinhold, New York (1976).
120. Saechtling, H., *International Plastics Handbook*, 2nd ed., Hanser, Munich (1987).
121. Brandrup, J., and Immergut, E. H. (eds.), *Polymer Handbook*, 3rd ed., John Wiley, New York (1989).
122. Cheremisinoff, N. P., *Handbook of Polymer Science and Technology*, Vols. 1–4, Marcel Dekker, New York (1989).
123. Chanda, M., and Roy, S. K., *Plastics Technology Handbook*, 2nd ed., Marcel Dekker, New York (1992).
124. English, L. K., Computers in plastics: Product design and processing, *Mater. Eng.*, pp. 30–34 (Mar. 1985).
125. Wiggins, M. U., Polymer Selection Assistant, Sample Knowledge Base, INSIGHT 2 Manual, Copyright © 1985, Level Five Research, Inc., 503 Fifth Avenue, Indialantic, FL 32903.
126. Lovrich, M. L., and Tucker, C. L., III, Computer program picks both resins and processes, *Plastics Eng.*, pp. 43–46 (Oct. 1986).
127. Shortliffe, E. H., *Computer-Based Medical Consultations: MYCIN*, American Elsevier, New York (1976).
128. Buchanan, B. G., and Shortliffe, E. H. (eds.), *Rule-Based Expert Systems: The MYCIN Experiments of the Stanford Heuristic Programming Project*, Addison-Wesley, Reading, MA (1985).
129. Weiss, S., and Kalikowaski, C., *A Practical Guide to Designing Expert Systems*, Rowman & Allanheld (1984).
130. Sell, P. S., *Expert Systems: A Practical Introduction*, John Wiley, New York (1985).
131. Forsyth, R. (ed.), *Expert Systems*, Chapman & Hall, London (1985).
132. Harmon, P., and King, D., *Expert Systems: Artificial Intelligence in Business*, John Wiley, New York (1985).
133. Waterman, D., *A Guide to Expert Systems*, Addison-Wesley, Reading, MA (1986).
134. Jackson, P., *Introduction to Expert Systems*, Addison-Wesley, Reading, MA, (1986).
135. Van Horn, M., *Understanding Expert Systems*, The Waite Group, Bantam Books, New York (1986).
136. Hart, A., *Knowledge Acquisition for Expert Systems*, McGraw-Hill, New York (1986).
137. Bonnet, A., Hatton, J. P., and Truong-Ngoc, J.-M., *Expert Systems*, Prentice-Hall, Englewood Cliffs, NJ (1988).

138. Harmon, P., Maus, R., and Morrissey, W., *Expert Systems: Tools and Applications*, John Wiley, New York (1988).
139. Siddall, J. N., *Expert Systems for Engineers*, Marcel Dekker, New York (1990).
140. Hayes-Roth, F., Waterman, D., and Lenat, D. (eds.), *Building Expert Systems*, Addison-Wesley, Reading, MA (1984).
141. Brownston, L., Farrell, R., and Kant, E., *Programming Expert Systems in OPS5: An Introduction to Rule-Based Programing*, Addison-Wesley, Reading, MA (1985).
142. Martin, J., and Oxman, S., *Building Expert Systems: A Tutorial*, Prentice-Hall, Englewood Cliffs, NJ (1988).
143. Ntuen, C. A., Step-by-step procedure provides a beginning in developing an expert system, *Ind. Eng.*, pp. 33–36 (Oct. 1991).
144. Stylianou, A. C., Madey, G. R., and Smith, R. D., Selection criteria for expert system shells: A socio-technical framework, *Commun. ACM*, 35: 31–48 (Oct. 1992).

Appendix A

Suppliers for Engineering Plastics

Product	Supplier
ABS	Borg-Warner Chemicals, Inc. International Center Parkersburg, WV 26101
	Mobay Chemical Corp. Penn Lincoln Parkway, West Pittsburgh, PA 15205
	Monsanto Co. 800 North Lindbergh Boulevard, A3NG St. Louis, MO 63167

Product	Supplier
	USS Chemicals, Division of U.S. Steel Corp. 600 Grant Street (28th Floor) Pittsburgh, PA 15230
Acetals	E. I. du Pont de Nemours & Co., Inc. 1007 Market Street Wilington, DE 19898
	Celanese Plastics and Specialties Co. 26 Main Street Chatham, NJ 07928
Nylon 6	Allied Corporation, Fibers and Plastics Co. P.O. Box 2332R Morristown, NJ 07960
	Adell Plastics, Inc. 4530 Annapolis Road Baltimore, MD 21227
	Custom Resins, Division of Bemis Co., Inc. P.O. Box 933 Henderson, KY 42420
	Nypel, Inc. 24 Union Hill Road West Conshohocken, PA 19428
Nylon 6/6	E. I. du Pont de Nemours & Co., Inc. 1007 Market Street Wilmington, DE 19898
	Celanese Plastics and Specialties Co. 26 Main Street Chatham, NJ 07928
	Adell Plastics, Inc. 4530 Annapolis Road Baltimore, MD 21227
	Monsanto Co. 800 North Lindbergh Boulevard, A3NG St. Louis, MO 63167
	Nypel, Inc. 24 Union Hill Road West Conshohocken, PA 19428
	A. Schulman, Inc. 3550 Market Street Akron, OH 44313

Product	Supplier
	Wellman, Inc.—Plastics Division 75 Federal Street Boston, MA 02110
Polycarbonate	General Electric Plastics Operation One Plastics Avenue Pittsfield, MA 01201
	Mobay Chemical Corp. Penn Lincoln Parkway, West Pittsburgh, PA 15205
PPE-based resins	General Electric Plastics Operation One Plastics Avenue Pittsfield, MA 01201
Polypropylene	Adell Plastics, Inc. 4530 Annapolis Road Baltimore, MD 21227
	Fiberfil Division, Dart Industries, Inc. P.O. Box 3333 Evansville, IN 47732
	Fiberite Corp. 515 West Third Street Winona, MN 55987
	Hercules, Inc. 910 Market Street Wilmington, DE 19899
	LNP Corp. 412 King Street Malvern, PA 19355
	Thermofil, Inc. 815 North 2nd Avenue Brighton, MI 48116
Polysulfone	Union Carbide Corp. Specialty Chemicals and Plastics Division Old Ridgebury Road Danbury, CT 06817
Polyesters	E. I. du Pont de Nemours & Co., Inc. 1007 Market Street Wilmington, DE 19898

Product	Supplier
	Celanese Plastics and Specialties Co. 26 Main Street Chatham, NJ 07928
	General Electric Plastics Operation One Plastics Avenue Pittsfield, MA 01201
	GAF Corp. 140 West 51st Street New York, NY 10020
Teflon® PPA	E. I. du Pont de Nemours & Co., Inc. 1007 Market Street Wilmington, DE 19898

Appendix B

Details of Various Files Needed for the Expert System SELECTHER

When using a formatted system diskette, besides the COMMAND.COM file there should be a total of 13 files as listed below.

SELECTA.BAT
SELECTA.HLP
SELECTA.OUT
SELECTA.RUL
SELECTA.TXT
EXSYS.EXE
EX1.HLP
EX2.HLP

EX3.HLP
EX4.HLP
EX5.HLP
EX6.HLP
REPORT

Details of the files are given below.

Contents of SELECTA.BAT

```
ECHO OFF
CLS
DEL REPORT
CLS
EXSYS SELECTA NOQUESTIONS RECOVER
CLS
: EXIT
```

Contents of SELECTA.HLP

~Q 1
In case you would like to guess the possible chemical environment that the thermoplastic would be exposed to, then you may choose any one or more of the proposed eight chemicals/solvents. Choosing none of the above or unknown will assume that the thermoplastic is not affected by any of the chemicals/solvents listed.

~Q 2
Extended exposure to elevated temperatures or below freezing temperatures both have detrimental effects on the structural properties of resins and complete failure of the part may occur in use.

~Q 3
At normal temperatures at around 73 degrees Fahrenheit, none of the thermoplastics, including PC and PET are affected greatly by immersion in water. However, at higher temperatures exceeding 125 degrees Fahrenheit, PC and PET are known to hydrolyze to such an extent that the physical properties losses are extensive.

~Q 4
Nylons gain or lose moisture based on the humidity changes in the atmosphere. If the average humidity is 50%, nylons absorb 2.5–3.0% water. When this happens, there usually are dimensional changes in the molded part and there will also be major changes in such structural properties as stiffness and strength.

~Q 5
The coefficient of expansion of metal and the thermoplastic, particularly of the unreinforced versions, are very different from each other. Hence, when there are temperature changes, the difference in expansion can cause part warpage or cracking.

~Q 6
A heavily loaded glass reinforced resin is not going to have the gloss of a resin without glass, and hence, if appearance is important, then the selection has to be limited to natural, unreinforced resins.

~Q 7
Tensile strength is the resistance of the material to being pulled apart. It is usually expressed in pounds per square inch (p.s.i.) or K p.s.i. which is equal to 1000 p.s.i.

~Q 8
Flexural modulus is the term relating to stiffness of the material and basically represents the force required to break a sample by bending or flexing. It is usually expressed in pounds per square inch (p.s.i.) or K p.s.i. which is equal to 1000 p.s.i.

~Q 9
Notched Izod test consists of breaking a test bar and calculating the amount of energy required to break it. It gives a measure of the impact resistance of the material, namely, its ability to withstand such mechanical abuse as being struck by a blow of a hammer or struck by a dropped weight.

~Q 10
Elongation at break is the property always associated with tensile strength and is basically the increase in original length of the material at the point of fracture when it is pulled apart. It is expressed as a percentage.

~Q 11
Creep (or deformation under load) may be defined as the deformation of a material that takes place over extended periods of time while the material is supporting a load. It is usually expressed in pounds per square inch (p.s.i.) or K p.s.i. which is equal to 1000 p.s.i.

~Q 12
Dielectric strength is the ratio of the dielectric breakdown voltage to the thickness of an insulating material. It is expressed in Volts/mil. Most resin suppliers provide the dielectric strength of their materials based on tests made on samples 125 mils thick.

~Q 13
Arc resistance is the time in seconds that an arc may play across the surface of a material without rendering it conductive.

Contents of SELECTA.OUT

```
FILE REPORT /A
"                              REPORT"
"                              --------------"
"  "
"************************************************************************
"  "
"USER INPUT"
"-----------------"
V2 /L
V3 /L
Q1 /L
Q2 /L
Q3 /L
Q4 /L
Q5 /L
Q6 /L
Q7 /L
Q8 /L
Q9 /L
Q10 /L
Q11 /L
Q12 /L
Q13 /L
"SELECTHER OUTPUT"
"-----------------------------"
C /L
V48 /L
V49 /L
V50 /L
V51 /L
"  "
"************************************************************************
```

Files SELECTA.RUL and SELECTA.TXT

The files SELECTA.RUL and SELECTA.TXT are basically the rule file and the text file, respectively, and are generated through the development editor EDITXS when the expert system is built. The information that is needed for the generation of these two files is given below. (Further details can be obtained from the EXSYS manual.)

QUALIFIERS:

1 The Engineering Thermoplastic product during end-use application is likely to be exposed to the following chemicals/solvents:

Strong acid
Strong base
Salt solution
Aromatics
Gas
Alcohol
Ketone
Aldehyde
None of the above
Unknown

2 The Engineering Thermoplastic product during end-use application is

held at all times at normal temperatures of about 73 degrees Fahrenheit
held frequently at elevated temperatures for extended period of time
held frequently at temperatures below freezing (less than 32 degrees Fahrenheit) for extended period of time

3 The Engineering Thermoplastic product during end-use application is at all times

immersed in water whose temperature exceeds 125 degrees Fahrenheit
immersed in water whose temperature is normally at around 73 degrees Fahrenheit
not immersed in water

4 The ambient relative humidity (where the product is being manufactured and/or being used) is

greater than 50% and changes frequently with time
greater than 50% and does not change with time
less than 50% and changes frequently with time
less than 50% and does not change with time

5 The proposed application involves the use of metals and the thermoplastic as part of a single item:

Yes
No

6 Good product appearance and gloss is

of critical importance
not of critical importance

7 The Tensile Strength (TS) value in K p.s.i. needs to be in the range of

4 <= (TS) < 23
7 <= (TS) < 23
10 <= (TS) < 23
15 <= (TS) < 23
19 <= (TS) < 23
21 <= (TS) < 23

8 The Flexural Modulus (FM) value in K p.s.i. needs to be in the range of

100 <= (FM) < 1400
300 <= (FM) < 1400
400 <= (FM) < 1400
600 <= (FM) < 1400
800 <= (FM) < 1400
1000 <= (FM) < 1400
1200 <= (FM) < 1400

9 The Notched Izod (NI) value at 73 degrees F in ft-lb/in needs to be in the range of

0.4 <= (NI) < 50
1.5 <= (NI) < 50
3 <= (NI) < 50
10 <= (NI) < 50
25 <= (NI) < 50

10 The Elongation at Break (EB) value in percentage needs to be in the range of

3 <= (EB) < 400
5 <= (EB) < 400
7 <= (EB) < 400
10 <= (EB) < 400
25 <= (EB) < 400
75 <= (EB) < 400
100 <= (EB) < 400
200 <= (EB) < 400
300 <= (EB) < 400

11 The Creep Modulus (CM) value in K p.s.i. needs to be in the range of

35 <= (CM) < 1400
50 <= (CM) < 1400
75 <= (CM) < 1400
150 <= (CM) < 1400
250 <= (CM) < 1400
350 <= (CM) < 1400
450 <= (CM) < 1400
550 <= (CM) < 1400
650 <= (CM) < 1400
750 <= (CM) < 1400
1000 <= (CM) < 1400
1200 <= (CM) < 1400

12 The Dielectric Strength (DS) value in V/mil needs to be in the range of

350 <= (DS) < 600
400 <= (DS) < 600
450 <= (DS) < 600
500 <= (DS) < 600
550 <= (DS) < 600

13 The Arc Resistance (AR) value in sec. needs to be in the range of

5 <= (AR) < 250
50 <= (AR) < 250
75 <= (AR) < 250
100 <= (AR) < 250
125 <= (AR) < 250
150 <= (AR) < 250

CHOICES:

1 ABS
2 Acetal
3 6 nylon
4 6/6 nylon
5 6/6 nylon (30% glass)
6 6/6 nylon (mineral reinforced)
7 6/6 nylon (Super Tough)
8 Polycarbonate
9 Polycarbonate (30% glass)
10 PPE-based resin
11 PPE-based resin (30% glass)
12 Polypropylene (40% talc)
13 Polysulfone
14 Polybutylene terephthalate (30% glass)
15 Polyethylene terephthalate (30% glass)

VARIABLES:

1 P_ABS
Price of ABS in cents per cubic inch
Numeric variable
Initialized to 3.400000

2 ECP
Estimated Cost of Product in cents per cubic inch
Numeric variable

3 MAX_ET
Maximum End-use Temperature in degrees Fahrenheit
Numeric variable

4 HDT_ABS
Heat Deflection Temperature of ABS at 264 p.s.i. in degrees Fahrenheit
Numeric variable
Initialized to 216.000000

5 ABS
ABS
Numeric variable
Initialized to 0.000000

6 P_ACETAL
Price of Acetal in cents per cubic inch
Numeric variable
Initialized to 6.300000

7 HDT_ACETAL
Heat Deflection Temperature of Acetal at 264 p.s.i. in degrees Fahrenheit
Numeric Variable
Initialized to 277.000000

8 ACETAL
Acetal
Numeric variable
Initialized to 0.000000

9 P_6N
Price of 6 Nylon in cents per cubic inch
Numeric variable
Initialized to 5.900000

10 HDT_6N
Heat Deflection Temperature of 6 Nylon at 264 p.s.i. in degrees Fahrenheit
Numeric variable
Initialized to 170.000000

11 6NYLON
6 Nylon
Numeric variable
Initialized to 0.000000

12 P_66N
Price of 6/6 Nylon in cents per cubic inch
Numeric variable
Initialized to 56.500000

13 HDT_66N
Heat Deflection Temperature of 6/6 Nylon at 264 .p.s.i. in degrees Fahrenheit
Numeric variable
Initialized to 194.000000

14 66NYLON
6/6 Nylon
Numeric variable
Initialized to 0.000000

15 P_66N30G
Price of 6/6 Nylon (30% glass) in cents per cubic inch
Numeric variable
Initialized to 9.800000

16 HDT_66N30G
Heat Deflection Temperature of 6/6 Nylon (30% glass) at 264 p.s.i. in degrees Fahrenheit

Numeric variable
Initialized to 485.000000

17 66NYLON30G
6/6 Nylon (30% glass)
Numeric variable
Initialized to 0.000000

18 P_66N_MR
Price of 6/6 Nylon (mineral reinforced) in cents per cubic inch
Numeric variable
Initialized to 6.300000

19 HDT_66N_MR
Heat Deflection Temperature of 6/6 Nylon (mineral reinforced) at 264 p.s.i. in degrees Fahrenheit
Numeric variable
Initialized to 446.000000

20 66NYLON_MR
6/6 Nylon (mineral reinforced)
Numeric variable
Initialized to 0.000000

21 P_66N_ST
Price of 6/6 Nylon (Super Tough) in cents per cubic inch
Numeric variable
Initialized to 7.700000

22 HDT_66N_ST
Heat Deflection Temperature of 6/6 Nylon (Super Tough) at 264 p.s.i. in degrees Fahrenheit
Numeric variable
Initialized to 159.000000

23 66NYLON_ST
6/6 Nylon (Super Tough)
Numeric variable
Initialized to 0.000000

24 P_PC
Price of Polycarbonate in cents per cubic inch
Numeric variable
Initialized to 6.700000

25 HDT_PC
Heat Deflection Temperature of Polycarbonate at 264 p.s.i. in degrees Fahrenheit
Numeric variable
Initialized to 270.000000

26 PC
Polycarbonate

Numeric variable
Initialized to 0.000000

27 P_PC30G
Price of Polycarbonate (30% glass) in cents per cubic inch
Numeric variable
Initialized to 10.400000

28 HDT_PC30G
Heat Deflection Temperature of Polycarbonate (30% glass) at 264 p.s.i. in degrees Fahrenheit
Numeric variable
Initialized to 295.000000

29 PC30G
Polycarbonate (30% glass)
Numeric variable
Initialized to 0.000000

30 P_PPE
Price of PPE-based resin in cents per cubic inch
Numeric variable
Initialized to 5.200000

31 HDT_PPE
Heat Deflection Temperature of PPE-based resin at 264 p.s.i. in degrees Fahrenheit
Numeric variable
Initialized to 265.000000

32 PPE
PPE-based resin
Numeric variable
Initialized to 0.000000

33 P_PPE30G
Price of PPE-based resin (30% glass) in cents per cubic inch
Numeric variable
Initialized to 10.700000

34 HDT_PPE30G
Heat Deflection Temperature of PPE-based resin (30% glass) at 264 p.s.i. in degrees Fahrenheit
Numeric variable
Initialized to 300.000000

35 PPE30G
PPE-based resin (30% glass)
Numeric variable
Initialized to 0.000000

36 P_PP40T
Price of Polypropylene (40% talc) in cents per cubic inch
Numeric variable
Initialized to 2.300000

37 HDT_PP40T
Heat Deflection Temperature of Polypropylene (40% talc) at 264 p.s.i. in degrees Fahrenheit
Numeric variable
Initialized to 170.000000

38 PP40T
Polypropylene (40% talc)
Numeric variable
Initialized to 0.000000

39 P_PPS
Price of Polysulfone in cents per cubic inch
Numeric variable
Initialized to 20.900000

40 HDT_PS
Heat Deflection Temperature of Polysulfone at 264 p.s.i. in degrees Fahrenheit
Numeric variable
Initialized to 345.000000

41 PS
Polysulfone
Numeric variable
Initialized to 0.000000

42 P_PBT30G
Price of Polybutylene terephthalate (30% glass) in cents per cubic inch
Numeric variable
Initialized to 10.000000

43 HDT_PBT30G
Heat Deflection Temperature of Polybutylene terephthalate (30% glass) at 264 p.s.i. in degrees Fahrenheit
Numeric variable
Initialized to 405.000000

44 PBT30G
Polybutylene terephthalate (30% glass)
Numeric variable
Initialized to 0.000000

45 P_PET30G
Price of Polyethylene terephthalate (30% glass) in cents per cubic inch
Numeric variable
Initialized to 7.400000

46 HDT_PET30G
Heat Deflection Temperature of Polyethylene terephthalate (30% glass) at 264 p.s.i. in degrees Fahrenheit
Numeric variable
Initialized to 435.000000

47 PET30G
Polyethylene terephthalate (30% glass)
Numeric variable
Initialized to 0.000000

48 CAUTION_1
If the product is held frequently at elevated temperatures for extended period of time, then please note that heat aging can cause eventual deterioration of polymer properties or even complete failure of its structure. It is important to bear this point in mind during resin selection.
Displayed at the end of a run as text only
Displayed at the end of a run

49 CAUTION_2
If the product is held frequently below freezing (less than 32 degrees Fahrenheit) for extended period of time, then please note that impact resistance and toughness are reduced drastically even by a drop of few degrees in temperature in this range. It is important to bear this point in mind during resin selection.
Displayed at the end of a run as text only
Displayed at the end of a run

50 CAUTION_3
If the proposed application involves the use of metals and the thermoplastic as part of a single item, then please note that care has to be taken to allow for differences in the coefficient of thermal expansion values between the two, when there is likelihood of temperature changes. It is important to bear this point in mind during resin selection and keep away from unreinforced resins if possible.
Displayed at the end of a run as text only
Displayed at the end of a run

51 MESSAGE_1
The resin selection process has failed to give you a choice. Probably the anticipated environmental and structural requirements are too demanding and restrictive. Kindly relax one or more of the requirements and rerun the program to see if it does give a possible list of choices for resin selection.
Displayed at the end of a run as text only
Displayed at the end of a run

File EXSYS.EXE

This file provides the inference engine to run the expert system and is a proprietary product of EXSYS, Inc., P.O. Box 75158, Contract Station 14, Albuquerque, NM 87194, USA.

Contents of EX1.HLP

A V S	SELECTHER	Help

If you wish to know why SELECTHER considers one of the IF conditions to be true, enter the line number of the IF condition. SELECTHER will respond by presenting the origin of the information:

1. If the information came from asking you for input, SELECTHER will let you know that "you told it" so.
2. If the condition that you are asking about has not been checked yet, SELECTHER will tell you that "it does not yet know if it is true".
3. If the statement is a mathematical expression, then SELECTHER will display the value of each variable in the expression. If you want to know how the value for a specific variable was determined, enter the variable line number. SELECTHER will explain how that value was got.

Note that SELECTHER will re-display the rule after each question.

Press the <ENTER> key when you have finished asking about the rule.

Contents of EX2.HLP

A V S	SELECTHER	Help

SELECTHER is basically asking you for a numerical input in order to proceed with the consultation. When the numbered list has more than one item, input the relevant number(s) separated by a comma or space and then press the <ENTER> key to make the selection. If there is just a single item, then input the number 1 and press the <ENTER> key to continue.

If you do not understand why SELECTHER is asking you this question, you can ask it what rule it is trying to apply by typing WHY and then pressing the <ENTER> key. SELECTHER will respond by displaying the rule(s) it is trying to validate. Pressing the <ENTER> key will re-display the original screen.

Typing <?> and pressing the <ENTER> key gets more details such as definitions of certain terms. However, this facility is not available for every screen.

If you want to store the data that you have input so far and exit SELECTHER, enter QUIT in response to the request for information. When asked to name the file in which the data is to be stored, DO NOT USE the name SELECTA as the data will get stored in existing file and previous information will be erased.

Contents of EX3.HLP

A V S	SELECTHER	Help

SELECTHER is basically asking you for a numerical input in order to proceed with the consultation. When the numbered list has more than one item, input the relevant number(s) separated by a comma or space and then press the <ENTER> key to make the selection. If there is just a single item, then input the number 1 and press the <ENTER> key to continue.

Since you are only changing input and not applying any rule, you cannot ask SELECTHER the question WHY for more information at this stage.

If you want to store the data you have input so far and exit SELECTHER, then enter QUIT in response to the request for information. You will be asked to name the file in which the data is to be stored. DO NOT USE the name SELECTA because when there is already a file with that name, the data will get stored in it and all previous information will be erased.

Contents of EX4.HLP

A V S	SELECTHER	Help

SELECTHER has displayed the RESINS along with their certainty factors based on requirements that were input by you. You may now ask SELECTHER about the rule(s) it fired to arrive at a particular answer by entering the line number.

Pressing <C> will display all inputs that SELECTHER acquired. By entering a line number, you can alter the data. When the data is in the form you wish, you may press <R> to rerun.

Pressing <P> will allow you to print the output screen in standard format or in the report generator form. In the standard format, you have the option of also printing all the input that was given to SELECTHER.

Pressing [Q] will allow you to store data and exit SELECTHER. When asked for the filename to store the data, DO NOT use the name SELECTA as the original information in the existing file will be erased during data storage.

Pressing <A> or <G> will re-display the list; pressing <D> ends consultation.

Contents of EX5.HLP

A V S	SELECTHER	Help

SELECTHER has displayed the RESINS along with their certainty factors based on requirements that were input by you. You may now ask SELECTHER about the rule(s) it fired to arrive at a particular answer by entering the line number.

Pressing <C> will display all inputs that SELECTHER acquired. By entering a line number, you can alter the data. When the data is in the form you wish, you may press <R> to rerun.

Pressing <P> will allow you to print the output screen in standard format or in the report generator form. In the standard format, you have the option of also printing all the input that was given to SELECTHER.

Pressing [Q] will allow you to store data and exit SELECTHER. When asked for the filename to store the data, DO NOT use the name SELECTA as the original information in the existing file will be erased during data storage.

Pressing <A> or <G> will re-display the list; pressing <D> ends consultation.

Contents of EX6.HLP

A V S	SELECTHER	Help

SELECTHER has displayed all the input it acquired during the program run. Input the line number that you wish to change. SELECTHER will re-ask you the appropriate question and then re-display the new list of input. When you have modified the input to the way you wish it to be, press <R> for a rerun.

If, due to the new data, SELECTHER needs to get additional information, you will be asked for it. So also if new resins are to be displayed, SELECTHER will show them to you. You can ask WHY when SELECTHER asks for input and you will be able to view the rules.

When all the new information has been sorted, SELECTHER will display the new resin list along with the certainty factors. You can continue changing the data and re-running the consultation as many times as you wish.

Once you have made changes, you can return to your original input by pressing <O> (the letter O, not zero).

Contents of REPORT

REPORT

****

USER INPUT

Estimated Cost of Product in cents per cubic inch = 25

Maximum End-use Temperature in degrees Fahrenheit = 100

The Engineering Thermoplastic product during end-use application is likely to be exposed to the following chemicals/solvents: None of the above

The Engineering Thermoplastic product during end-use application is held at all times at normal temperatures of about 73 degrees Fahrenheit

The Engineering Thermoplastic product during end-use application is at all times not immersed in water

The ambient relative humidity (where the product is being manufactured and/or being used) is less than 50% and does not change with time

The proposed application involves the use of metals and the thermoplastic as part of a single item: No

Good product appearance and gloss is not of critical importance

The Tensile Strength (TS) value in K p.s.i. needs to be in the range of 4 <= (TS) < 23

The Flexural Modulus (FM) value in K p.s.i. needs to be in the range of 100 <= (FM) < 1400

The Notched Izod (NI) value at 73 degrees F in ft-lb/in needs to be in the range of 0.4 <= (NI) < 50

The Elongation at Break (EB) value in percentage needs to be in the range of 3 <= (EB) < 400

The Creep Modulus (CM) value in K p.s.i. needs to be in the range of 35 <= (CM) < 1400

The Dielectric Strength (DS) value in V/mil needs to be in the range of 350 <= (DS) < 600

The Arc Resistance (AR) value in sec. needs to be in the range of 5 <= (AR) < 250

SELECTHER OUTPUT

Polyethylene terephthalate (30% glass): Probability=71/100

Polybutylene terephthalate (30% glass): Probability=62/100

6/6 nylon (30% glass): Probability=60/100

Acetal: Probability=59/100

PPE-based resin: Probability=59/100

6/6 nylon (mineral reinforced): Probability=58/100

PPE-based resin (30% glass): Probability=58/100

Polycarbonate (30% glass): Probability=55/100

6/6 nylon: Probability=50/100

Polypropylene (40% talc): Probability=50/100

Polysulfone: Probability=50/100

ABS: Probability=48/100

Polycarbonate: Probability=47/100

6 nylon: Probability=44/100

6/6 nylon (Super Tough): Probability=41/100

****

Glossary

Amorphous polymer A polymer that has no crystalline component and in which there is no order or pattern to the distribution of the molecules.

Apparent viscosity The ratio of shear stress to shear rate which has not been corrected for entrance-length effects in a capillary rheometer.

Arc resistance The time in seconds that an arc may play across the surface of a material without rendering it conductive.

Artificial intelligence (AI) The area of computer science and cognitive psychology which deals with the development of intelligent behavior of machines. This includes machines/computer programs that solve problems from experience (expert systems), understand language (natural-language processing systems), interpret visual scenes (vision systems), and perform motor functions (robots).

Backward chaining A method of inferencing within an expert system which begins with what is to be proven and attempts to establish facts required to prove it.

Clamping force The force required to prevent mold opening during injection molding.

Compressive strength The ability of a material to resist forces that tend to crush or compress it.

Creep (or deformation under load) The deformation of a material that takes place over extended periods of time while the material is supporting a load.

Crystalline polymer A polymer that has an ordered structural arrangement of molecules.

Declarative knowledge Information required by an expert system which is known through facts, theories, rules, hypotheses, or heuristics.

Dielectric strength The ratio of the dielectric breakdown voltage to the thickness of an insulating material.

Distortion The deformation resulting from differential shrinkage occurring due to the relaxation of residual stresses formed because of nonuniform solidification in an injection-molded product.

Elongation at break The increase in original length of a material at the point of fracture when it is pulled apart, expressed as a percentage. This property is always associated with tensile strength.

Environmental stress cracking The name given to a phenomenon by which a plastic product under high stresses may crack when in contact with certain active environments, such as detergents, fats, and silicone fluids.

Expert system A highly proficient computer-based problem solver which embodies specialized knowledge in a narrow, specific area of practical importance.

Fatigue endurance The ability of a material to resist fatigue failure.

Fatigue failure The failure of a material when it is subjected repeatedly to stresses below its elastic limit.

Flash The thin excess web of material that is forced into crevices between mating mold surfaces.

Flexural modulus A term relating to stiffness of a material, which basically represents the force required to break a sample by bending or flexing.

Flexural strength The ability of a material to resist forces that tend to bend it.

Flow activation energy The energy required to activate viscous flow.

Flow curve or rheogram A curve relating shear stress or viscosity to shear rate.

Forward chaining A method of inferencing within an expert system in which the IF portions of the rules are matched against known facts in order to establish new facts.

Freeze-off temperature The temperature at which a polymeric melt solidifies in the mold.

Glass transition temperature The temperature at which increased molecular mobility results in significant change in properties.

Heuristics Rules of thumb which may not guarantee solutions but which assist in narrowing the possible alternatives in order to reach acceptable solutions.

Impact resistance The ability of a material to resist forces that tend to break it when it is dropped or struck by a sharp blow.

Inference engine The portion of an expert system which processes the knowledge base and uses deductive reasoning to reach conclusions.

Izod impact A test for shock loading in which a notched specimen held at one end is broken by striking and the energy absorbed is measured.

Knowledge base The portion of an expert system which houses the domain knowledge.

Knowledge representation The structuring of domain knowledge in such a way as to make it easy to build an expert system.

Melt flow index (MFI) The weight of a polymer in grams that is extruded in 10 min through a capillary of specific diameter and length by pressure applied through dead weight under prescribed temperature conditions as per set international standards.

Melt flow indexer The apparatus used for measuring melt flow index.

Model An idealized relationship of behavior expressible in mathematical terms.

Normal stress coefficient The ratio of the normal stress to the square of the shear rate.

Normal stress difference The difference between normal stress components.

Power-law model Behavior characterized by a power (n) relationship between shear stress and shear rate.

Procedural knowledge Provides instructions in terms of strategies and tactics on how to use the declarative knowledge in an expert system.

Production rule In an expert system, formal representation in the form of IF-THEN statements which parallels the knowledge representation in the long-term memory of a human being.

Relaxation time The time for a stress to decrease to an exponential inverse of its initial value under constant strain.

Rheogram or flow curve A curve relating shear stress or viscosity to shear rate.

Stress cracking The appearance of minute cracks when a material is stressed, thereby leaving a site which is particularly vulnerable to physical and chemical attack.

Tensile strength The ability of a material to withstand forces tending to pull it apart.

Tensile strength at yield The strength corresponding to the transition from elastic to plastic deformation.

Ultimate elongation The amount of stretch that a material can undergo before taking a permanent set or otherwise failing.

Ultimate tensile strength The maximum attainable load acting on a tensile test specimen divided by the original cross-sectional area of the specimen.

User interface The portion of an expert system which permits two-way communication between the end user and the expert system.

Yield stress The stress corresponding to the transition from elastic to viscous deformation of flow.

Nomenclature

Symbol	Description	Units	Eq.
a_H	Heat diffusion coefficient of melt	—	(5.3)
$\overline{A}$	Effective projected area of molding	cm^2	(5.4)
b	Sample width	in.	(3.1), (3.3)
$\overline{B}$	Numerical function dependent on n	—	(5.5), (5.6)
$\overline{C}$	Proportionality constant	—	(5.1), (5.3)
d	Sample thickness	in.	(3.1), (3.3)
D	Deflection	in.	(3.1), (3.3)
E	Activation energy for viscous flow of polymer melt (Table 4.3 for values)	kcal/mol	(4.10)
E_{p1}	Value of E for polymer P_1 in the preparation of polyblends	kcal/mol	(4.17)
E_{P2}	Value of E for polymer P_2 in the preparation of polyblends	kcal/mol	(4.17)

Symbol	Description	Units	Eq.
E^*	Modulus	psi	(3.4)
F	Force exerted by the test load L on the polymer in the melt flow indexer	dynes	(4.1)
K	Consistency index whose values are tabulated in Tables 4.6 and 4.7	$(g/cm \cdot s^{2-n}) \cdot (g/10\ min)^n$	(4.13), (4.14), (4.16)
K_{P1}	Consistency index for polymer 1 in the blend	$(g/cm \cdot s^{2-n}) \cdot (g/10\ min)^n$	(4.18)
K_{P2}	Consistency index for polymer 2 in the blend	$(g/cm \cdot s^{2-n}) \cdot (g/10\ min)^n$	(4.18)
l	Length of span between supports	in.	(3.1)–(3.3)
l_N	Length of nozzle	cm	(4.1)
L	Test load, i.e., dead weight + piston weight	kg	(4.3), (4.6)
L_1	Test load 1	kg	(4.8)
L_2	Test load 2	kg	(4.8)
MFI	Melt flow index	g/10 min	(4.4), (4.6), (4.7)
MFI_{L1}	Melt flow index determined under test load 1	g/10 min	(4.8)
MFI_{L2}	Melt flow index determined under test load 2	g/10 min	(4.8)
MFI_{T1}	Melt flow index at first temperature T_1	g/10 min	(4.9), (4.10)
MFI_{T2}	Melt flow index at second temperature T_2	g/10 min	(4.9), (4.10)
$MFI_{P1,TP1}$	Melt flow index of polymer 1 at temperature T_{P1} in blends	g/10 min	(4.17)
$MFI_{P2,TP2}$	Melt flow index of polymer 2 at temperature T_{P2} in blends	g/10 min	(4.17)
MFI_{B,T_c}	Melt flow index of blend at compounding temperature T_c	g/10 min	(4.18), (4.19)
n	Power-law index whose values are given in Tables 4.6 and 4.7	—	(4.13), (4.14), (4.16)
n_{P1}	Power-law index for polymer 1 in the blend	—	(4.18)
n_{P2}	Power-law index for polymer 2 in the blend	—	(4.18)
N	Power index in the Carreau model	—	(4.11)
N_{P1}	Power index for polymer 1 in the blend	—	(4.19)
N_{P2}	Power index for polymer 2 in the blend	—	(4.19)
P	Power index in the general rheological model whose values are given in Table 4.7	—	(4.14)
$P_{0,min}$	Minimum value of P_0	$dynes/cm^2$	(5.1), (5.4)
P_0	Pressure drop from the center of a "spreading disk" flow in the mold cavity during injection molding	$dynes/cm^2$	—
P^*	Load	lb	(3.1), (3.3)
Q	Volumetric flow rate	cm^3/s	(4.2)
R	Gas constant = 1.9858	cal/mol·K	(4.10), (4.17)
R_d	Instantaneous radius of the disk or flow length in the mold during injection molding	cm	(5.1)
R_N	Radius of nozzle	cm	(4.1), (4.2)
R_P	Radius of piston	cm	(4.1)
S	Stress	psi	(3.1), (3.3)
T	Polymer melt temperaturex	°C	(5.3)

Symbol	Description	Units	Eq.
T_c	Compounding temperature	°C	(4.17)
T_g	Glass transition temperature of polymer	K	(4.9)
T_s	Standard reference temperature, equal to $T_g + 50$	K	(4.9)
T_0	Freeze-off temperature in the mold during injection molding	°C	(5.3)
T_1	ASTM recommended test temperature	K	(4.9), (4.10)
T_2	Temperature at which MFI is required	K	(4.9), (4.10)
T_{P1}	Temperature of MFI measurement for polymer 1	°C	(4.17)
T_{P2}	Temperature of MFI measurement for polymer 2	°C	(4.17)
x_0	Thickness of the original cavity of the mold in injection molding	cm	(5.1)
α^1	Ellis model parameter	—	(4.12)
γ	Strain	—	(3.2)
$\dot{\gamma}$	Shear rate	s^{-1}	(4.2), (4.4), (4.7), (4.11), (4.16)
$\dot{\gamma}_c$	Compounding shear rate	s^{-1}	(4.18)
η	Steady shear viscosity	poise	(4.5), (4.6), (4.11), (4.15)
η_0	Zero shear viscosity	poise	(4.11)–(4.14)
θ_m	Mold temperature	°C	(5.5)
λ	Time constant and model parameter	s	(4.11)
ρ	Polymer melt density	g/cm^3	(4.4), (4.6), (4.7)
σ	Stress (Figures 3.22–3.27)	N/m^2	
σ_0	Stress value at an elapsed time of 1 day (Figures 3.23, 3.25, 3.27)	N/m^2	
τ	Shear stress	$dynes/cm^2$	(4.1), (4.3), (4.5), (4.15), (4.16)
$\tau_{1/2}$	Special value of shear stress when steady shear viscosity is half the zero shear viscosity	$dynes/cm^2$	(4.12)

Index